INTRODUCTION TO THE THEORY OF THE RAMAN EFFECT

INTRODUCTION TO THE THEORY OF THE RAMAN EFFECT

by

J. A. KONINGSTEIN
Carleton University, Ottawa, Canada

D. REIDEL PUBLISHING COMPANY

DORDRECHT-HOLLAND

Library of Congress Catalog Card Number 72–77876

ISBN 90 277 0276 4

Printed in The Netherlands by D. Reidel, Dordrecht

to M.G.

INTRODUCTION

This book is written particularly for chemists. Being one myself, I have on several occasions tried to find a book where the theory of molecular Raman spectroscopy is treated, and not being able to find one which satisfactorily answered the questions I wanted to see answered, I decided to try to write a book on it myself.

Back in the middle fifties I was shown a Raman spectrum for the first time: some faint lines on a photographic plate. In the fall of 1971, during a visit to Moscow, I vividly remembered that spectrum when the son of Mandel'shtam showed me the first spectrum taken in Russia by his father and Landsberg in 1928. The spectrum of quartz photographed during January and February of that year showed the presence of some faint new lines and in later exposures these lines became stronger and stronger. Dr Mandel'shtam told me that his father had asked Landsberg to carry out more experiments to make sure that these lines, which they could not observe in the spectrum of the mercury lamp, were genuine. Originally they were looking for what we now call a Brillouin spectrum of quartz and the rather large shifts observed were not expected.

Sir C. V. Raman, who died when this book was being written, became interested in light scattering of liquids during a trip to Europe in 1921. Travelling by steamer he became fascinated by the colour of the Mediterranean Sea and back in India he commenced light scattering experiments with liquids. In the course of these experiments, which he carried out together with Krishnan and Venkatesvaran, frequency shifted lines were observed in the spectrum of liquid benzene on 28 February, 1928. The discovery was published in *Nature* (31 March and 21 April, 1928) while the Mandel'shtam and Landsberg article appeared in *Die Naturwissenschaften*. Their publication concluded with the following remark: "At this moment (6 May, 1928), it is not possible for us to judge if, and to what extent, there exists a relation between the phenomenon observed by us and that for the first time shortly discussed by Raman, because of the brevity of his description". They thus acknowledged Raman's work.

In actual fact, the Austrian physicist Smekal had already in 1923 predicted the possibility of the occurrence of frequency shifted lines, but Raman was not aware of his work. Also in 1928, Cabannes and Rocard in France tried to detect frequency shifted lines in the spectrum of gases, but they gave up their attempts.

The basis of this book can be found in Placzek's paper in Marx's *Handbuch der Radiologie*. There he explained the details of the molecular Raman effect, and using his results it is easy to find the formulas for the hyper, stimulated (and inverse), and electronic Raman effect.

However, a period of more than thirty years elapsed before these processes became more familiar to us. While early as 1930 Rasetti had already found an electronic Raman effect and Stoicheff had tried to observe induced absorption before the advent of the laser, the first stimulated Raman spectrum was excited and observed by Woodbury and Ng in 1962. Stoicheff and Jones observed the inverse Raman spectrum in 1963, and the hyper Raman effect was found in 1965 by Terhune, Maker, and Savage in 1965. Laser excited electronic Raman effects were observed in 1966. The frustrations and disappointments encountered by who knows how many scientists trying to get spectra of solids during the pre-laser period were legio. Many of the difficulties have now been removed – one has only to look at the results obtained by Porto and co-workers in the late sixties on single crystals.

This book provides background information on the theory of molecular Raman spectroscopy. I have tried for instance to introduce the irreducible tensor components and their relevance to electronic and rotational Raman spectroscopy. In order to keep the book to a reasonable size I have assumed of the reader some knowledge of group theory and quantum mechanics. At the same time, the reader should not expect to find Raman spectra of compounds, frequencies of fundamental vibrations, etc., but he may be able to handle the Raman activity of the lines of the various Raman processes, even if it comes to employing double groups.

I am greatly indebted to O. Sonnich Mortensen for his work on the irreducible tensors, and I have also benefited from discussions with H. J. Bernstein and W. F. Murphy and G. Herzberg who kindly lent me the book containing Placzek's paper.

Finally I wish to thank Mrs A. Ratz for typing and correcting the manuscript. I of course take full responsibility for the contents of the book and for the fact that the language is not quite perfect. Its preparation took up much of my time, which I otherwise would have spent with my family.

Ottawa and Paris, 1971

TABLE OF CONTENTS

CHAPTER I

INTERACTION OF LIGHT AND MATTER, THE RAMAN SCATTERING TENSOR

Light can interact in different ways with matter. It can be absorbed and emitted, but it may also be scattered. While the processes which govern these types of interactions are quite different, yet many aspects of the former types are of importance in understanding the scattering process. Using classical theories, it can be shown that an oscillating dipole – be it an induced or a permanent one – generates electromagnetic radiation. The radiation fields of dipoles, quadrupoles or magnetic dipoles are not equal and, as an example of one of the aspects mentioned above, may illustrate the fact that the spatial distribution of Raman scattered radiation and that of an oscillating dipole, quadrupole or magnetic dipole are closely related to each other.

At the heart of the mathematical formulation of Rayleigh and Raman scattering lies the expression derived for the scattering tensor. Although appropriate formulae were derived by Kramers and Heisenberg before the advent of the general acceptance of Quantum mechanics, there is also the method in which Quantum mechanics is used to describe the position of electrons of atoms or molecules – and energy – but for the introduction of the perturbation of the presence of electromagnetic radiation, classical expressions are employed. This method is described as the correspondence principle, in contrast to other calculations where the radiation field is treated Quantum mechanically and the scattering particle and the field are considered as one system. In the former method, the *density* of the radiation field of the incident and Raman scattered light do not appear. Einstein showed that absorption and emission of radiation can be either spontaneous or induced (or stimulated), and in the expressions for the absorption and emission processes, the density of the radiation field appears. Consequently, the correspondence principle theory falls short of explaining the stimulated and inverse Raman process, though it permits the calculation of the induced dipole moment between two quantum states of the scattering particle. This, in turn, provides information on the spatial distribution of the scattered light.

I-1. Electromagnetic Radiation of an Oscillating Dipole

According to the theory of Hertz, the relation between the total intensity (I) per second of the radiation field of an oscillating dipole is:

$$I = \frac{2}{3c^2} \overline{\ddot{M}} \qquad \text{(I, 1-1)}$$

where $\overline{\ddot{M}^2}$ stands for the time average of the second derivative of the dipole moment.

The total intensity (I) refers to the intensity emitted in all directions in space and it is also understood that the moment **M** is averaged over all orientations in space of oscillating dipoles of there is more than one dipole. In order to clarify (I, 1-1) it is assumed that a particle is situated at the origin of a Cartesian coordinate system and that the system is exposed to electromagnetic radiation travelling towards the particle along the x-axis with a velocity c. If ν_0 is the frequency of the plane-polarized wave, then the strength of the electric field at a distance x is:

$$\mathbf{E} = \mathbf{E}_z = E_0 \cos(2\pi\nu_0 t - x/c). \qquad \text{(I, 1-2)}$$

If it is now assumed that the field **E** induces an oscillating dipole moment which is in phase with and codirectional to the incident wave vector, it follows that:

$$M = \alpha E_0 \cos(2\pi\nu_0 t - x/c) \qquad \text{(I, 1-3)}$$

and

$$\ddot{M}^2 = 16\pi^4\nu_0^4\alpha^2 E_0^2 (\cos^2 2\pi\nu_0 t - x/c) \qquad \text{(I, 1-4)}$$

thus

$$\overline{\ddot{M}^2} = 8\pi^4\nu_0^4\alpha^2 E_0^2 \qquad \text{(I, 1-5)}$$

where α is the polarizability of the system.

A combination of Equations (I, 1-1) and (I, 1-5) yields for the total radiation emitted in space:

$$I = \frac{16\pi^4\nu_0^4}{3c^3}\alpha^2 E_0^2 . \qquad \text{(I, 1-6)}$$

The radiation density (flux per cm^2) of the incident light of linear polarization is:

$$I_0 = \frac{c}{4\pi} E_z^2 = \frac{c}{8\pi} E_0^2 \qquad \text{(I, 1-7a)}$$

and total electromagnetic flux, radiated in the solid angle of 4π radians is then:

$$I = \frac{128\pi^5\nu_0^4}{3c^4}\alpha^2 I_0 . \qquad \text{(I, 1-7b)}$$

Equation (I, 1-7b), however, does not yield information on the energy density and flux density in any specific direction in space.

The classical radiation theory of an oscillating dipole is in fact based on the description of the electromagnetic field produced by an accelerated charge employing Maxwell's equation. If the dipole moment is defined as $\mathbf{p} = e\mathbf{r}$ and assuming that the positive (or negative) charge executes linear harmonic motion with a frequency ω then, $\mathbf{p} = \mathbf{p}_0 \cos\omega t$ and the absolute value of

$$|\mathbf{E}| = |\mathbf{H}| = \frac{\omega^2}{r_0 c^2} p \sin\theta \qquad \text{(I, 1-8)}$$

where r_0 is the distance of the accelerated charge and a point in space and θ is the angle between $\mathbf{r}_0$ and the direction of the dipole moment **p** (Figure I-1,2). The radiation field of the linear harmonic oscillating electrical charge is such that only linearly-polarized waves emerge and the *amplitude* varies as the sine of the angle θ. Furthermore, the value of $|\mathbf{E}|$ is maximum for $\theta = 90°$ and if the dipole is surrounded by a sphere, see Figure I-2, then the vector **E** can be represented as being along the tangent to a meridian drawn on this sphere (and the vector **H** will be along the tangent to the corresponding parallel). The energy density ϱ of an electromagnetic field is given by:

$$\varrho = \frac{1}{8\pi}(\mathbf{E}^2 + \mathbf{H}^2) = \frac{1}{4\pi}\mathbf{E}^2 \qquad \text{(I, 1-9)}$$

and thus:

$$\varrho = \frac{1}{4\pi}\frac{\omega^4}{r_0^2 c^4} p^2 \sin^2\theta . \qquad \text{(I, 1-10)}$$

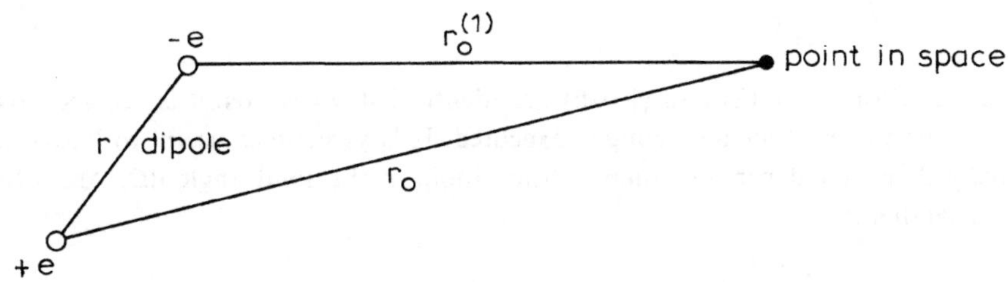

Fig. I-1. The position of a dipole in space.

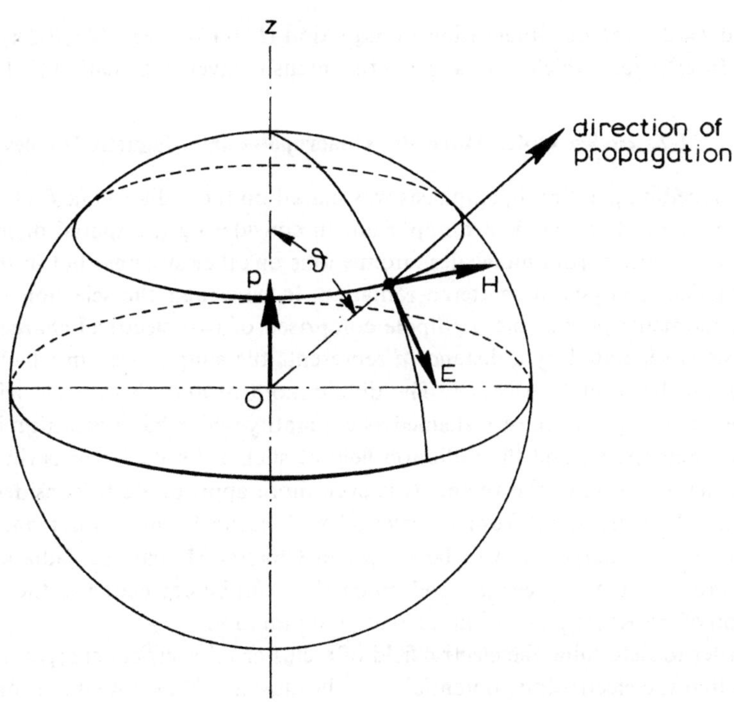

Fig. I-2. Polarization of the radiation field of an electric dipole.

The electromagnetic flux density $I(\theta)$ (where $I(\theta)=\varrho c$) in a given direction defined by the angle θ, is then:

$$I(\theta) = \frac{\omega^4 p^2}{4\pi r_0^2 c^3} \sin^2 \theta . \qquad \text{(I, 1-11)}$$

Equation (I, 1-11) represents the spatial intensity distribution of the oscillating dipole moment (Figure I-4). The intensity is a minimum in the direction of the dipole and assumes maximum values in directions perpendicular to the dipole. The total radiation I emitted into space is obtained by integrating (I, 1-11) over a sphere with radius r_0. The resulting expression is:

$$I = \frac{\omega^4 p^2}{3c^3} . \qquad \text{(I, 1-12)}$$

The Equations (I, 1-12) and (I, 1-6) are identical if $p=p_0 \cos\omega t$ is replaced by αE_0 and if the proper time averaging is executed. It is sometimes useful to know the intensity $\mathrm{d}I$ radiated per unit time by the dipole in the solid angle $\mathrm{d}\Omega$. According to classical theory:

$$\mathrm{d}I = \frac{\omega^4 p^2}{4\pi c^3} \sin^2 \theta \, \mathrm{d}\Omega \qquad \text{(I, 1-13)}$$

where $\mathrm{d}\Omega = 2\pi \sin\theta \, \mathrm{d}\theta$. Integration of Equation (I, 1-13) over $\theta (0 \leqslant \theta \leqslant \pi)$ yields the result: $I=\omega^4 p^2/3c^3$ which is the same as the intensity given in Equation (I, 1-12).

I-2. Higher Order Moments: Quadrupoles and Magnetic Dipoles

In the preceding paragraph, emphasis was placed on the radiation field of an induced dipole moment. Such fields are important in considering the spatial distribution of scattered radiation from molecules, atoms, ions or other systems; but in the detailed mathematical analysis of scattered radiation, in particular the selection rules, other electric moments play a role. A dipole comprised of two electrical charges with opposite signs separated by a distance d represents the simpler way in which electrical charges can be brought together. The dipole moment and also the electric field are considered vectors. A vector is defined as a quantity which has a direction in space as well as a magnitude, and the radiation field of such a dipole vector is thus identical zero in the direction of the vector. It is even more appropriate to consider electrical moments which result not from a cluster of two electrical charges but from clusters of many electrical charges. It will be shown in Chapter II that the radiation field of higher order moments – electric and magnetic – can be correlated to the spatial distribution of the Rayleigh and Raman scattered radiation.

In order to determine the electric field of a cluster of electrical charges, it is important to find the electrostatic potential V of the cluster. If this potential is measured at a distance r from the electrical charges where r is larger than the distances of the charges, then, according to Poisson:

$$\nabla^2 V = 0. \tag{I, 2-1}$$

∇^2 is the Laplacian operator and defined as:

$$\nabla^2 = \frac{\partial^2}{\partial x^2} + \frac{\partial^2}{\partial y^2} + \frac{\partial^2}{\partial z^2}. \tag{I, 2-2}$$

If the potential and the Laplacian operator are related to the polar coordinates r, θ, φ then the following expression is obtained:

$$\frac{1}{r^2}\frac{\partial}{\partial r}\left(r^2 \frac{\partial V}{\partial r}\right) + \frac{1}{r^2 \sin^2\theta}\left(\sin\theta \frac{\partial V}{\partial \theta}\right) + \frac{1}{r^2 \sin\theta}\frac{\partial^2 V}{\partial \varphi^2} = 0. \tag{I, 2-3}$$

If V is separated into a function $R(r)$ and $f(\theta, \varphi)$, where the former denotes the dependence of the distance r and the latter the angular dependence, then solutions of the partial differential equation are:

$$\frac{\partial}{\partial r}\left(r^2 \frac{\partial R}{\partial r}\right) - l(l+1)R = 0 \tag{I, 2-4a}$$

and

$$\frac{1}{\sin\theta}\frac{\partial}{\partial \theta}\left(\sin\theta \frac{\partial f}{\partial \theta}\right) + \frac{1}{\sin^2\theta}\frac{\partial^2 f}{\partial \varphi^2} + l(l+1) f = 0. \tag{I, 2-4b}$$

Under the proper conditions – that a solution to the first Equation (I, 2-4a) should be finite at large values of r – an expression for $R(r)$ is:

$$R(r) = \frac{A}{r^{l+1}} \tag{I, 2-4c}$$

and A is an arbitrary constant. The Equation (I, 2-4b) is a familiar one and is encountered in all textbooks which deal with Quantum Mechanics and its application to the hydrogen atom. The solution to this equation is given in terms of the Legendre polynomials $P_l(\cos\theta)$ where:

$$P_l(\cos\theta) = \frac{1}{2^l l!}\frac{\partial^l (\cos^2\theta - 1)}{\partial \cos\theta^2} \tag{I, 2-4d}$$

and the general solution of Poisson's equation may be written in terms of the Cartesian coordinates. Realizing that the atomic wave functions p_x, p_y and p_z may be written in terms of the coordinates x, y and z respectively, the expression for V becomes:

$$\begin{aligned} V = \frac{A_0}{r} &+ \frac{1}{r^2}(A_{1x}p_x + A_{1y}p_y + A_{1z}p_z) + \\ &+ \frac{1}{r^3}(A_{2z^2}d_{z^2} + A_{2xz}d_{xz} + A_{2xy}d_{xy} + A_{2yz}d_{zy} + A_{2(x^2-y^2)}d_{x^2-y^2}) + \\ &+ \frac{1}{r^4}\quad(\text{sum of } 4f \text{ orbitals}). \end{aligned} \tag{I, 2-5}$$

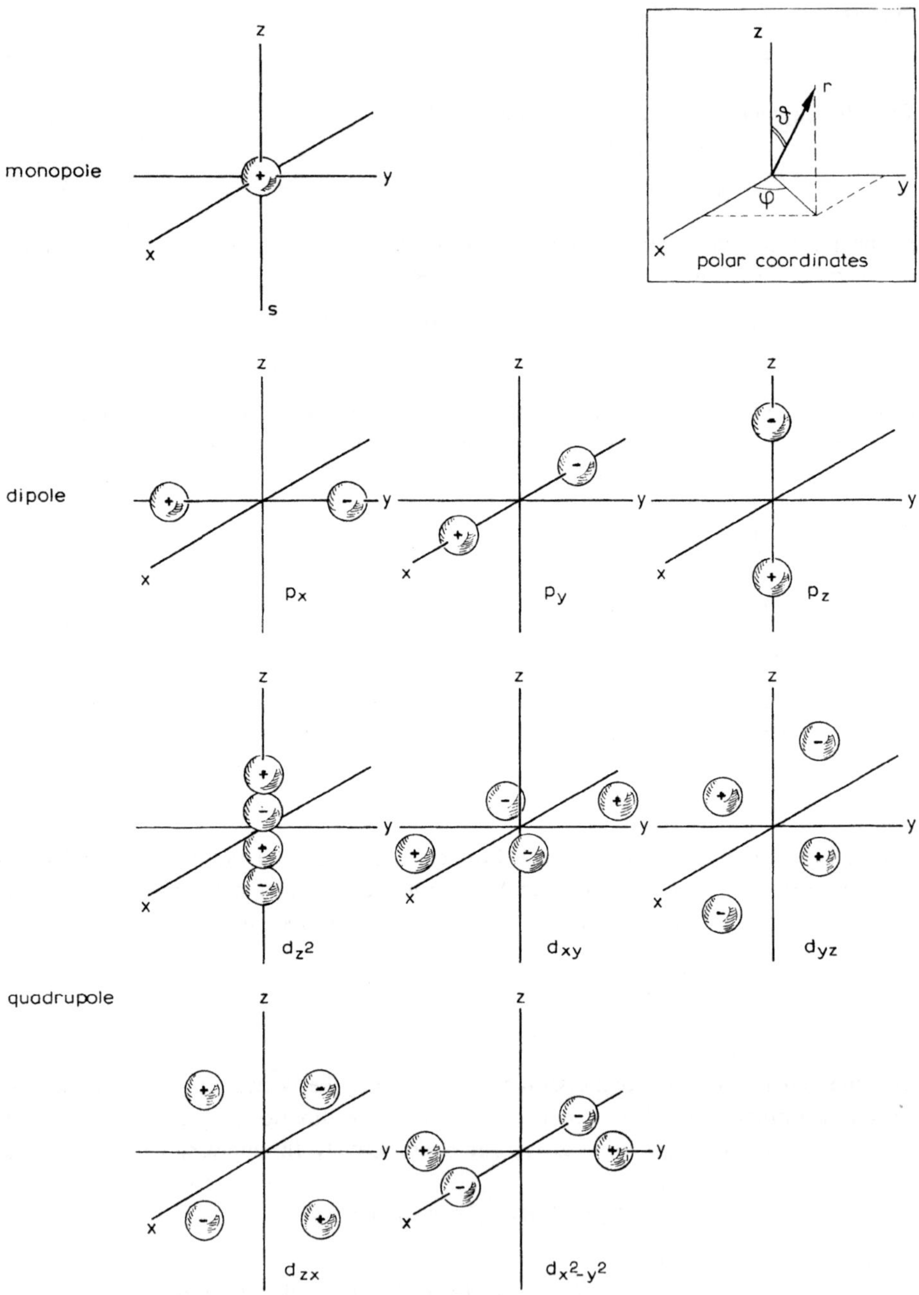

Fig. I-3. Pictorial representation of a monopole, dipole and quadrupole. Also indicated is the comparison with the (approximate) charge distribution of *s*, *p* and *d* orbitals.

The term A_0 equals the net charge of the group of electrical charges, but the first member of Equation (I, 2-5) is least sensitive to the distance r. For sufficiently large values of r it may thus be said that the group of electrical charges behaves as a single charged point. The second member represents contributions of the dipole moments where the electrical charges appear in pairs and the line joining the charges is either in the x, y or z direction of the Cartesian coordinate system. The dipole moments are shown in Figure I-3 and as a result of the fact that other terms of Equation (I, 2-5) dependent on $1/r^l$, the normal dipole moments are important if the total net charge of the cluster equals zero. It is also possible to make some definite remarks on the spatial distribution of the electrostatic field. This field has the same spatial orientation as the well-known p_x, p_y and p_z orbitals (Figure I-4).

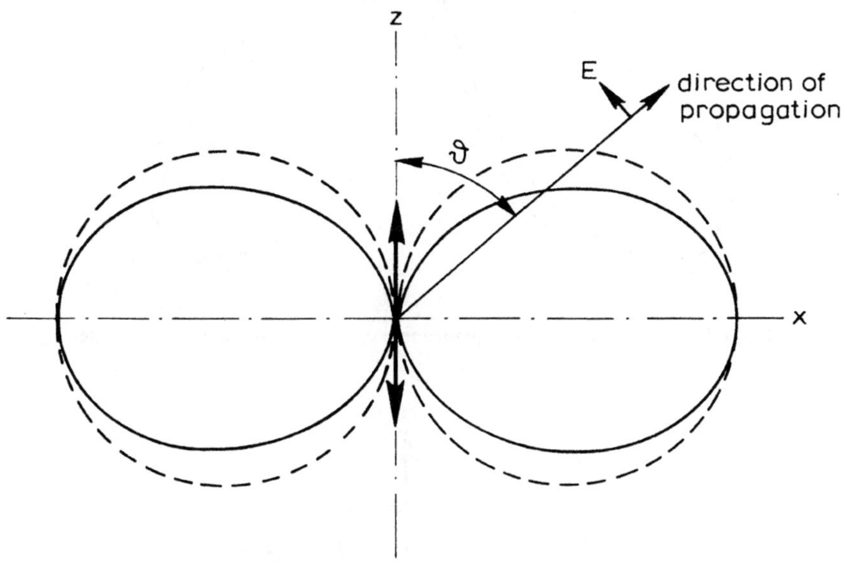

Fig. I-4. Radiation field of an electric dipole: - - - amplitude, ——— intensity.

The third member of Equation (I, 2-5) represents contribution of the socalled quadrupoles. The arrangement of the electrical charges is in pairs and this higher order moment is depicted in Figure I-3. There are two pairs, but the combination of the pairs results in five independent moments. The electrical field associated with these charges is similar to the way the 5 d orbitals are oriented in space. An example is given in Figure I-5. Apart from the different spatial dependence of the fields of dipoles and quadrupoles, the former may be represented by a vector, but this is not so for the quadrupole moment. A vector has three components – along the x, y or z-axis – but the quadrupole has five different components and is not a vector. A proper representation of the quadrupole moment is a dyad or a tensor. The reader is referred to Chapter II for a discussion of tensors. Higher order moments such as octupoles, etc., are not of

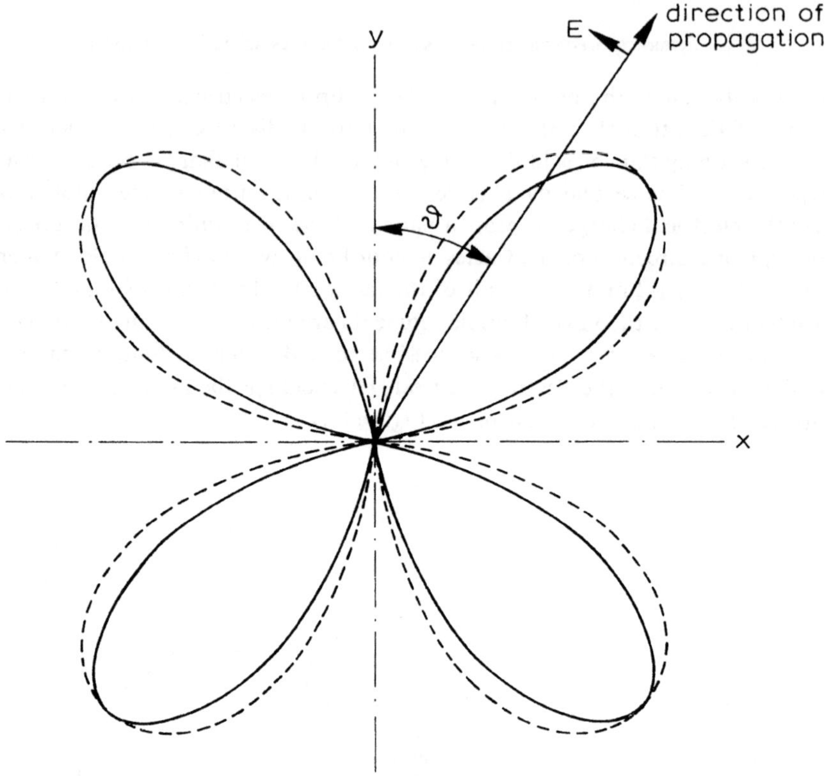

Fig. I-5. Radiation field of a quadrupole d_{xy}: - - - amplitude, ——— intensity.

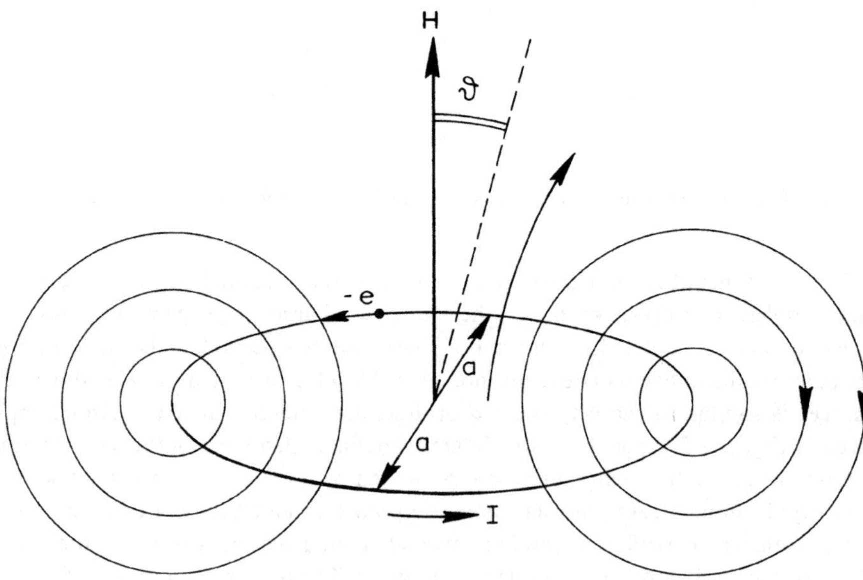

Fig. I-6. Model of an elementary magnetic dipole. The magnitude of the **E** and **H** vectors as function of the angle ϑ is given by Equation (I, 2-12).

importance here. However, electromagnetic fields are also associated with oscillating magnetic dipoles. A short discussion of magnetic dipoles is thus appropriate.

If an electric current I travels through a wire, then this current gives rise to a magnetic field. If the wire is bent in a ring, then the magnetic field is that depicted in Figure I-6. According to Ampere, the magnetic fields of this ring may be compared to the field of a uniformly magnetized disc if the area of ring and disc are equal. If an observer is far away from the disc or ring, then the magnetic field appears to be associated with a point dipole. The magnetic moment $\boldsymbol{\mu}$ of the ring is equal to:

$$|\boldsymbol{\mu}| = \text{current} \times \text{area}. \qquad \text{(I, 2-6)}$$

If the radius of the ring is a, and c is the velocity of light, then the expression for μ in electromagnetic units is:

$$|\boldsymbol{\mu}| = \pi I \frac{a^2}{c}. \qquad \text{(I, 2-7)}$$

For an electron with mass m_e moving in a ring of radius a with a velocity of v, the angular momentum **M** is defined as $\mathbf{M}=m_e va$; and for an electron, this moment is equal to:

$$|\mathbf{M}| = m_e va = \sqrt{l(l+1)}\,\hbar \qquad \text{(I, 2-8a)}$$

where l is the angular momentum quantum number. On the other hand, the electrical charge $-e$, moving with the velocity v passes a given point of the ring $v/2\pi a$ times per second so that the current is:

$$I = \frac{-ev}{2\pi a} \qquad \text{(I, 2-8b)}$$

and the magnetic moment is thus equal to:

$$|\boldsymbol{\mu}| = \frac{-eva}{2c}. \qquad \text{(I, 2-8c)}$$

A combination of Equations (I, 2-8a) and (I, 2-8c) gives:

$$|\boldsymbol{\mu}| = \frac{-e\hbar}{2m_e c}\sqrt{l(l+1)} = \frac{-e}{2m_e c}\mathbf{M}. \qquad \text{(I, 2-9)}$$

The axis of the magnetic moment $\boldsymbol{\mu}$ is antiparallel to the angular momentum **M** because of the negative charge of the electron. A classical picture of the angular momentum **M** is shown in Figure I-7. The momentum is perpendicular to the ring and both the radius **a** of the ring and the linear momentum $\mathbf{p}=m_e v$ may be considered vectors. The total angular momentum is $\mathbf{M}=\mathbf{a}\times\mathbf{p}$ where the product refers to the cross product of vectors. Such products are considered in detail in Chapter II. The radiation field of an oscillating magnetic dipole may be discussed in the following way. Suppose an oscillating magnetic dipole is that as depicted in Figure I-6. The electron follows only part of the orbit of Figure I-6, and the displacement of this electron may be written as $x=x_0 \cos 2\pi \nu_0 t$. The velocity $v=x'=-2\pi\nu_0 x_0 \sin 2\pi\nu t$ and ν_0 is the frequency of oscillation of the electron. From Equation (I, 2-9) it follows that the magnetic dipole moment will be:

$$\boldsymbol{\mu} = \boldsymbol{\mu}_0 \sin 2\pi\nu_0 t \tag{I, 2-10}$$

where

$$|\boldsymbol{\mu}_0| = \frac{e x_0 \nu_0 a}{c}. \tag{I, 2-11}$$

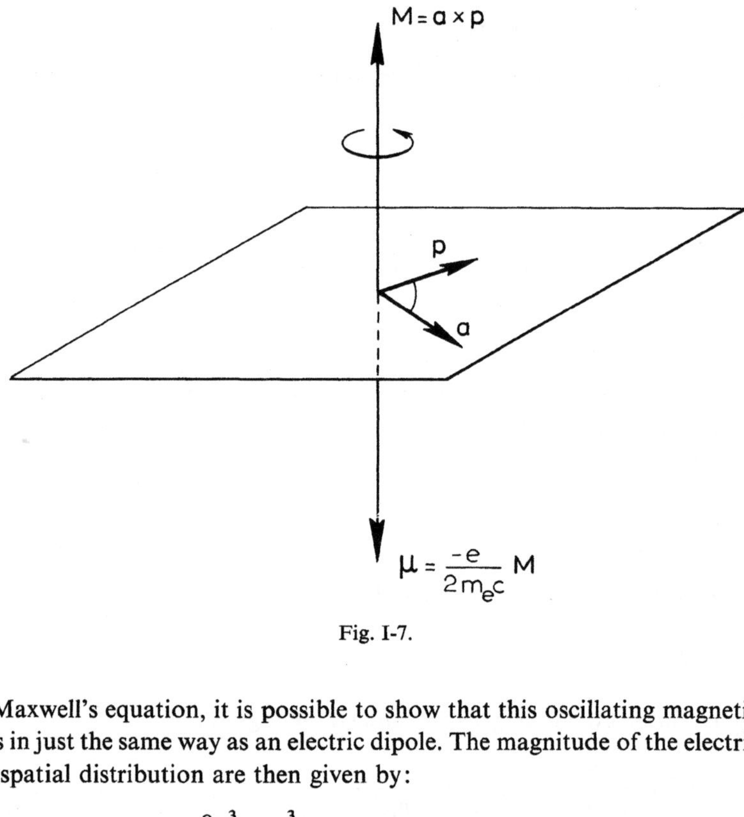

Fig. I-7.

Using Maxwell's equation, it is possible to show that this oscillating magnetic dipole radiates in just the same way as an electric dipole. The magnitude of the electric field **E** and its spatial distribution are then given by:

$$|\mathbf{E}| = |\mathbf{H}| = \frac{-8\pi^3 e x_0 \nu_0^3 a}{r c^3} \sin\theta \sin 2\pi\nu_0 (t - r/c). \tag{I, 2-12}$$

I-3. Spectroscopic Transitions

Up to this point, emphasis has been placed on the properties of classical radiators. However, all spectroscopic measurements involve transitions between states, and here an attempt is made to correlate results of classical theories with physical realities.

All spectroscopic measurements depend on the presence of some type of interaction between the atom, ion or molecule and an electromagnetic wave which can be regarded as an oscillating electric and magnetic field propagating through space. Before the particle can absorb energy from an electromagnetic wave, there must be some means by which it can interact with either the electric or magnetic fields. In order to interact with the electric field, the particle must have a charge distribution which changes when the system makes the transition from the initial to the final state – i.e. there must be an interaction of the instantaneous dipole moment in the particle with the electric field of the light wave. These transitions are called electric dipole transitions and a similar argument can be made for quadrupole and magnetic dipole transitions. In the latter case, the interaction of the instantaneous magnetic moment and the electromagnetic field of the wave occurs. This type of interaction is smaller than that of the electric dipole and electric field. From time-dependent Quantum mechanical calculations, it is found that the transitions between the states k and n with the wave functions ψ_k and ψ_n, depend on quantities called Einstein transition probabilities. Consider the general case of the states k and n with energies W_k and W_n; let only one wave function belong to these states (so that the levels are nondegenerate). There are N_k particles with energy W_k, and N_n particles with energy W_n. This system is assumed to be capable of acquiring energy from an external electromagnetic field so as to pass from k to n, or emit energy as electromagnetic radiation so as to pass from n to k (Figure I-8). The Einstein transition probabilities are then used to describe the following processes:

(i) *Spontaneous Emission.* This process is indicated in Figure I-8b and is related to the number of transitions of $n \rightarrow k$ per second. Thus:

$$-\frac{\mathrm{d}N_n}{\mathrm{d}t} = N_n A_{n \rightarrow k} \qquad \text{(I, 3-1)}$$

where $A_{n \rightarrow k}$ is the Einstein transition probability for spontaneous emission, and the quantity represents the probability that a system in state n will spontaneously emit energy and pass to the lower-lying state k.

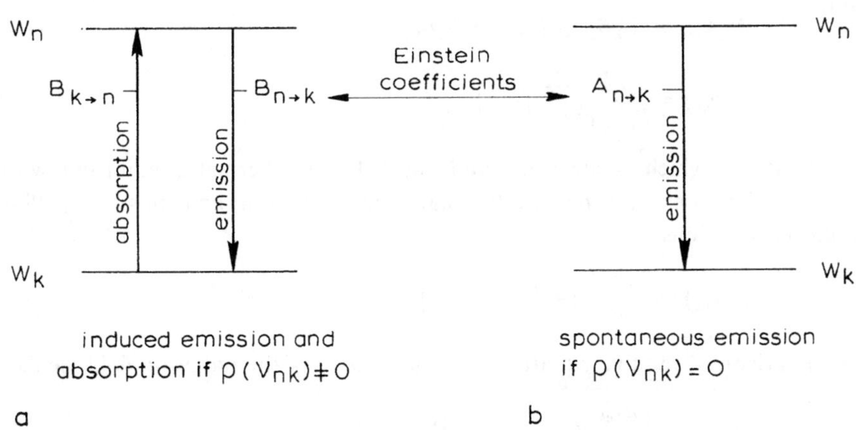

Fig. I-8.

(ii) *Induced (Stimulated) Absorption.* This process is related to the number of transitions of $k \to n$ (per second). Thus:

$$-\frac{\mathrm{d}N_k}{\mathrm{d}t} = N_k B_{k \to n} \varrho(\nu_{nk}) \tag{I, 3-2}$$

where $B_{k \to n}$ is the Einstein transition probability of induced absorption and $\varrho(\nu_{nk})$ is the radiation density due to electromagnetic radiation of energy $h(\nu_n - \nu_k)$.

(iii) *Induced (Stimulated) Emission.* A radiation field of density $\varrho(\nu_{nk})$ also induces the passage of the system from the upper state n to the lower state k accompanied by the emission of radiation of energy $h\nu_{nk}$. For the rate of emission it is found:

$$-\frac{\mathrm{d}N_n}{\mathrm{d}t} = N_n B_{n \to k} \varrho(\nu_{nk}) \tag{I, 3-3}$$

where $B_{n \to k}$ is the Einstein transition probability for induced (stimulated) emission. The radiation field is always present. Following Planck's laws for blackbody radiation, such a radiation field can be thought to exist of oscillators and even at absolute zero $\varrho(\nu) \neq 0$ because of the zero point energies of the oscillators.

The three transition probability coefficients $A_k \to_n$, $B_k \to_n$ and $B_{n \to k}$ are fundamental measures of the interaction of the particle with electromagnetic radiation. They are independent whether or not the systems are in a state of thermodynamic equilibrium. However, in order to obtain a relationship of the three quantities, it becomes necessary to consider the case of thermodynamic equilibrium. Boltzmann's distribution law is then applicable. Thus:

$$\frac{N_n}{N_k} = \frac{e^{-E_n/kT}}{e^{-E_k/kT}} = e^{-h\nu_{nk}/kT} \tag{I, 3-4}$$

and at equilibrium:

rate of emission = rate of absorption

Thus:

$$N_n A_{n \to k} + N_n B_{n \to k} \varrho(\nu_{nk}) = N_k B_{k \to n} \varrho(\nu_{nk}) \tag{I, 3-5a}$$

or:

$$\varrho(\nu_{nk}) = \frac{A_{n \to k}}{B_{k \to n}\, e^{h\nu_{nk}/kT} - B_{n \to k}}. \tag{I, 3-5b}$$

The radiation which is absorbed and emitted is in thermal equilibrium with the particles. This is exactly the case for blackbody radiation, and thus, using Planck's radiation equations:

$$\varrho(\nu_{nk}) = \frac{8\pi h \nu_{nk}^3}{c^3} \left[e^{-h\nu_{nk}/kT} - 1\right]^{-1}. \tag{I, 3-6}$$

A comparison of the two equations for the density of the radiation field yields:

$$A_{n \to k} = \frac{8\pi h \nu_{nk}^3}{c^3} B_{k \to n} \quad \text{and} \quad B_{k \to n} = B_{n \to k}. \tag{I, 3-7}$$

In the above, it is assumed that the particle has a permanent electric dipole moment. If e_j is the charge of the jth electron of the molecule, then the dipole moment in the x direction can be written as:

$$\mu_x = \sum_j e_j x_j \tag{I, 3-8}$$

where x_j is the projection of the distances of charges on the x-axis. In Quantum Mechanics, or, more particularly, in the situation where the molecule passes from state k to n, the average value of the dipole moment between the states should be nonvanishing; thus:

$$(\mu_x)_{kn} = \int \psi_n^* \sum_j e_j x_j \psi_k \, \mathrm{d}\tau \neq 0 \,. \tag{I, 3-9}$$

Intuitively it may be seen that the transition probabilities should be related to the (average) value of the dipole moment. Using the results of the Quantum Mechanical treatment, the relation is:

$$B_{n\to k} = B_{k\to n} = \frac{2\pi}{3\hbar^2} \left|(\mu_x)_{kn}\right|^2 . \tag{I, 3-10}$$

The ratio of the rate of spontaneously n_{sp} and induced (stimulated) n_{st} emitted radiation is obtained using Equations (I, 3-5) and (I, 3-10). The result is:

$$\frac{n_{\mathrm{sp}}}{n_{\mathrm{st}}} = \frac{N_k - 1}{N_n} = e^{h\nu_{nk}/kT} - 1 \,. \tag{I, 3-11}$$

Thus, in general, $n_{\mathrm{sp}} > n_{\mathrm{st}}$; however, situations may arise so that $N_n > N_k$ (population inversion), and then the rate of stimulated emission is larger than that for spontaneous emitted radiation. Presently there are many systems for which population inversion has been achieved and this, coupled with techniques to build up the intensity of stimulated light, are encountered in present-day masers and lasers.

I-4. Classical Consideration of the Radiation Field and Quantum Mechanical Calculation of the Induced Dipole Moment, the Correspondence Principle

In Section 1 of this chapter the radiation field associated with oscillating dipoles is discussed. This dipole moment may be permanent or induced but in discussing the scattering of light of molecules, atoms or other systems, the latter moment is paramount. The time-dependent wave function $\Psi^0(q, t)$ of electrons of the entity which shall be exposed to a bombardment of photons is, according to Schrödinger:

$$\mathscr{H}^0 \Psi^0(q, t) = -\frac{\hbar}{i} \frac{\partial}{\partial t} \Psi^0(q, t) \,. \tag{I, 4-1}$$

$\mathscr{H}^0$ is the Hamiltonian and $\Psi^0(q, t)$ refers to the wave function (q=coordinate, t=time) of the system prior to the onset of the interaction with the light waves. The general solution of Equation (I, 4-1) is obtained by introducing functions which depend on coordinates and time only, so that:

$$\Psi^0(q, t) = \psi(q)\,\varphi(t)\,. \qquad \text{(I, 4-2)}$$

Inclusion of the new functions in Equation (I, 4-1) does permit a separation of the Schrödinger equation. For the time-dependent part:

$$\frac{\mathrm{d}\varphi(t)}{\mathrm{d}t} = -\frac{i}{\hbar}\,W\varphi(t)$$

the solutions are:

$$\varphi_n(t) = e^{-i/\hbar\, W_n t} \qquad \text{(I, 4-3)}$$

where W_n specifies various values of the energy of the system. Similar expressions as (I, 4-1), (I, 4-2) and (I, 4-3) can be written to investigate the solutions of the complex conjungate function $\Psi^{*0}(q, t)$.

The part of the Schrödinger equation which relates to the function $\psi(q)$ yields also only solutions for specific values of W_n so that:

$$\mathscr{H}^0\psi_n(q) = W_n\psi_n(q)\,. \qquad \text{(I, 4-4)}$$

Thus, for specific values of n there are particular solutions to the differential Equations (I, 4-3) and (I, 4-4), and consequently, any linear combination will then also be a solution. Mathematically, these solutions may then be formulated as:

$$\Psi^0(q, t) = \sum_n a_n\psi_n(q)\, e^{-i/\hbar\, W_n t}\,. \qquad \text{(I, 4-5)}$$

Although Equation (I, 4-5) describes the situation for any unperturbed system, other expressions are necessary to describe the wave function of electrons of molecules, atoms or ions if these particles are exposed to electromagnetic radiation as is the case in Rayleigh and Raman spectroscopy.

If nonpolarized electromagnetic radiation is impinging on the system, then the electric vector **E** can be written as:

$$\mathbf{E} = \mathbf{A}^*\, e^{2\pi i\nu_0 t} + \mathbf{A}\, e^{-2\pi i\nu_0 t}\,. \qquad \text{(I, 4-6)}$$

ν_0 is the frequency of the light wave and **A** is a complex vector which can be written as $\mathbf{A} = \mathbf{a} + i\mathbf{b}$. Thus Equation (I, 4-6) can also be written as:

$$\mathbf{E} = 2\,(\mathbf{a}\cos 2\pi\nu_0 t + \mathbf{b}\sin 2\pi\nu_0 t)\,. \qquad \text{(I, 4-7)}$$

The vector **E** is always real and the absolute value of $|\mathbf{E}^2|$, where **E** is the time-averaged value, is given by:

$$|\mathbf{E}|^2 = 2\,(\mathbf{a}^2 + \mathbf{b}^2) = 2\,|\mathbf{A}|^2 \qquad \text{(I, 4-8)}$$

The radiation density is then (see Equation (I, 1-7a)):

$$I_0 = \frac{c}{4\pi}\,|\mathbf{E}|^2 = \frac{c}{2\pi}\,|\mathbf{A}|^2\,. \qquad \text{(I, 4-9a)}$$

For linear-polarized radiation travelling along the x-axis, so that the electric field is parallel to the z-axis:

$$\mathbf{E}_x = \mathbf{E}_y = 0; \quad \mathbf{A}_x = \mathbf{A}_y = 0 \quad \text{and} \quad \mathbf{E}_z = \mathbf{E}_0 \cos(2\pi\nu_0 t - \varphi)$$

so that $\mathbf{A}_z = \mathbf{E}_0/2\, e^{i\varphi}$. Thus also

$$\mathbf{E}_z = \mathbf{E}_0\, e^{2\pi i\nu_0 t - i\varphi} + \mathbf{E}_0\, e^{-2\pi i\nu_0 t + i\varphi}$$

and

$$|\mathbf{A}^2| = \mathbf{A}_z\mathbf{A}_z^* = \frac{E_0^2}{4}$$

from which follows that:

$$I_0 = \frac{c}{8\pi} E_0^2. \qquad \text{(I, 4-9b)}$$

Here φ is a phase factor which mathematically provides a method to fix any value of **E** at the origin of the Cartesian coordinate system. For circular-polarized radiation propagating along the x-axis, the expressions are:

$$\mathbf{E}_z = \mathbf{E}_0 \cos(2\pi\nu_0 t - \varphi); \quad \mathbf{E}_y = \mathbf{E}_0 \sin(2\pi\nu_0 t - \varphi); \quad \mathbf{E}_x = 0$$

so that

$$\mathbf{A}_z = i\mathbf{A}_y = \tfrac{1}{2}\mathbf{E}_0\, e^{i\varphi}; \quad \mathbf{E}_x = 0. \qquad \text{(I, 4-9c)}$$

Here again the radiation density I_0 is calculated as: $I_0 = c/4\pi\, E_0^2$. Natural light propagating along x can be represented by contributions from left- and right-handed polarized light so that:

$$\begin{aligned} &\mathbf{E}_z = \mathbf{E}_0 \cos(2\pi\nu_0 t - \varphi_1) \quad \text{and} \quad \mathbf{E}_y = \mathbf{E}_0 \sin(2\pi\nu_0 t - \varphi_2) \\ &\mathbf{E}_x = 0. \end{aligned} \qquad \text{(I, 4-9d)}$$

Obviously for natural light there does not exist a relation between the phase factors φ_1 and φ_2. If the incident radiation – be it plane-polarized, circular-polarized or natural light – impinges on the system of molecules, atoms or ions then the electric field **E** produces a potential at the place of the electrons of the particle. This perturbing potential can be written as:

$$\mathscr{H}^1 = -ME \quad \text{where} \quad M = \sum_j e_j \mathbf{r}_j \qquad \text{(I, 4-10)}$$

(e_j is the electrical charge of the jth electron and $\mathbf{r}_j$ is a position vector). The time-dependent Schrödinger equation of the perturbed system takes now the form:

$$(\mathscr{H}^0 - ME)\, \Psi^0(q, t) = -\frac{\hbar}{i} \frac{\partial \Psi^0}{\partial t}(q, t). \qquad \text{(I, 4-11)}$$

In first order perturbation calculations the wave function $\Psi_k(q, t)$ of a state k is written as:

$$\Psi_k(q, t) = \Psi_k^0(q, t) + \Psi_k^1(q, t). \qquad \text{(I, 4-12)}$$

Substitution of (I, 4-12) in (I, 4-11) and making use of (I, 4-2) results in

$$\left(\mathscr{H}^0 - \frac{\hbar}{i}\frac{\partial}{\partial t}\right) \Psi_k^1(q, t) = (EM)\, \Psi_k^0(q, t) \tag{I, 4-13}$$

and the solution of this partial differential equation is

$$\Psi_k^1(q, t) = \psi_k^+(q)\, e^{-i/\hbar\,(W_k + h\nu_0)t} + \psi_k^-(q)\, e^{-i/\hbar\,(W_k - h\nu_0)t}. \tag{I, 4-14}$$

A combination of Equations (I, 4-14), (I, 4-6) and the expression of the wave function of the state k given by $\Psi_k(q, t) = \psi_k^0(q)\, e^{-i/\hbar W_k t}$ can now be employed to generate a set of new expressions. In collecting from these expressions the coefficients of equal time dependence, the following relations are obtained.

$$\mathscr{H}^0\psi_k^+(q) - (W_k + h\nu_0)\,\psi_k^+(q) = (\mathbf{AM})\,\psi_k(q) \tag{I, 4-15}$$

$$\mathscr{H}^0\psi_k^-(q) - (W_k - h\nu_0)\,\psi_k^-(q) = (\mathbf{A}^*\mathbf{M})\,\psi_k(q). \tag{I, 4-16}$$

The RHS of (I, 4-15) and (I, 4-16) can also be written similar to

$$(\mathbf{AM})\,\psi_k(q) = \sum_r (\mathbf{AM}_{rk})\,\psi_r$$

where

$$\mathbf{M}_{kr} = \int \psi_r^* M \psi_k \,\mathrm{d}r. \tag{I, 4-16a}$$

The functions ψ_r are the solutions of the time-independent Schrödinger equation

$$\mathscr{H}^0\psi_r = W_r\psi_r$$

and a combination of (I, 4-13) and (I, 4-15), (I, 4-16) permits the determination of the functions $\psi_k^\pm$ in terms of the unperturbed functions ψ_r:

$$\psi_k^+ = \sum_r \frac{(\mathbf{AM}_{kr})}{W_r - W_k - h\nu_0}\,\psi_r$$

and

$$\psi_k^- = \sum_r \frac{(\mathbf{A}^*\mathbf{M}_{kr})}{W_r - W_k + h\nu_0}\,\psi_r. \tag{I, 4-17}$$

The wave function ψ_k^1 is then given by:

$$\Psi_k^1 = \frac{1}{h}\sum_r \psi_r \left[\frac{(\mathbf{AM}_{kr})}{\nu_{rk} - \nu_0}\, e^{-i/\hbar(W_k + h\nu_0)t} + \frac{(\mathbf{A}^*\mathbf{M}_{kr})}{\nu_{rk} + \nu_0}\, e^{-i/\hbar(W_k - h\nu_0)t}\right]. \tag{I, 4-18}$$

The expectation value of the moment M_{kk} in the situation that the particle is exposed to the electric field vector $\mathbf{E}$ is:

$$\mathbf{M}_{kk} = \int (\Psi_k^0 + \Psi_k^1)^*\, M\,(\Psi_k^0 + \Psi_k^1)\,\mathrm{d}\tau. \tag{I, 4-19}$$

After substituting (I, 4-18) and a similar expression for the complex conjugate wave function ψ_k^{1*} where

$$\Psi_k^{1*} = \frac{1}{h} \sum \psi_r^* \left[\frac{(\mathbf{A}^* M_{kr})}{\nu_{rk} - \nu_0} e^{i/h(W_k + h\nu_0)t} + \frac{(\mathbf{AM}_{kr})}{\nu_{rk} + \nu_0} e^{i/\hbar(W_k - h\nu_0)t} \right] \quad \text{(I, 4-20)}$$

the moment of the state k becomes:

$$\mathbf{M}_{kk} = \mathbf{M}_{kk}^0 + \frac{1}{h} \sum \left[\frac{(\mathbf{AM}_{kr})\, \mathbf{M}_{rk}}{\nu_{rk} - \nu_0} + \frac{\mathbf{M}_{kr}\, (\mathbf{AM}_{rk})}{\nu_{rk} + \nu_0} \right] e^{-2\pi i \nu_0 t} + \\ + \frac{1}{h} \sum_r \left[\frac{(\mathbf{A}^*\mathbf{M}_{kr})\, \mathbf{M}_{rk}}{\nu_{rk} + \nu_o} + \frac{\mathbf{M}_{kr}\, (\mathbf{A}^*\mathbf{M}_{rk})}{\nu_{rk} - \nu_0} \right] e^{2\pi i \nu_0 t}. \quad \text{(I, 4-21)}$$

where

$$\mathbf{M}_{kk}^0 = \int \Psi_k^0 M\, \Psi_k^0\, \mathrm{d}\tau .$$

Apparently $\mathbf{M}_{kk}^0$ is not dependent on time and not associated with radiation. After replacing M with $\sum_j e_j \mathbf{r}_j$ it is seen that this moment represents the permanent moment of the particle in the state k. The second part of the RHS of Equation (I, 4-21) can also be written as:

$$\mathbf{M}_{kk}^{(1)} = \mathbf{C}_{kk}\, e^{-2\pi i \nu_0 t} + \mathbf{C}_{kk}^*\, e^{2\pi i \nu_0 t}$$

where

$$\mathbf{C}_{kk} = \frac{1}{h} \sum_r \left[\frac{(\mathbf{AM}_{kr})\, \mathbf{M}_{rk}}{\nu_{rk} - \nu_0} + \frac{\mathbf{M}_{kr}\, (\mathbf{AM}_{rk})}{\nu_{rk} + \nu_0} \right]. \quad \text{(I, 4-22)}$$

The most interesting aspects of Equation (I, 4-22) are that the moment $\mathbf{M}_{kk}^{(1)}$ is real and that its time-dependent part is identical to that of the incident electromagnetic radiation. Also, if the incident light happens to be plane-polarized so that $\mathbf{A}$ of Equation (I, 4-21) is given by $\mathbf{A}_z = \frac{1}{2} E_0 e^{i\varphi}$ (see Equation (I, 4-9b)), then the phase of incident and scattered light is not changed. The polarization of this radiation is thus colinear and also in phase – and of the same frequency – with the incident radiation and is called coherent Rayleigh scattering. For the expectation value of the induced moment $\mathbf{M}_{kn}$ of the states k and n, the following expressions are obtained:

$$\mathbf{M}_{kn} = \int (\Psi_n^0 + \Psi_n^1)^*\, M\, (\Psi_k^0 + \Psi_k^0)\, \mathrm{d}\tau \\ = \mathbf{M}_{kn}^{(0)} + \mathbf{M}_{kn}^{(1)} + \mathbf{M}_{kn}^{(2)} \quad \text{(I, 4-23)}$$

where:

$$\mathbf{M}_{kn}^{(0)} = \mathbf{M}_{kn} e^{-2\pi i \nu_{kn} t} \quad \text{(I, 4-23a)}$$

$$\mathbf{M}_{kn}^{(1)} = \frac{1}{h} \sum_r \left[\frac{(\mathbf{AM}_{kr})\, \mathbf{M}_{rn}}{\nu_{rk} - \nu_0} + \frac{\mathbf{M}_{kr}\, (\mathbf{AM}_{rn})}{\nu_{rn} + \nu_0} \right] e^{-2\pi i (\nu_{kn} + \nu_0) t} + \quad \text{(I, 4-23b)}$$

$$+ \frac{1}{h} \sum_r \left[\frac{(\mathbf{A}^*\mathbf{M}_{kr})\, \mathbf{M}_{rn}}{\nu_{rk} + \nu_0} + \frac{\mathbf{M}_{kr}\, (\mathbf{A}^*\mathbf{M}_{rn})}{\nu_{rn} - \nu_0} \right] e^{-2\pi i (\nu_{kn} - \nu_0) t} \quad \text{(I, 4-23c)}$$

$$\mathbf{M}_{kn}^{(2)} = \frac{1}{h^2} \sum_{r,p} \left[\frac{(\mathbf{A}^*\mathbf{M}_{kp})\,\mathbf{M}_{pr}\,(\mathbf{A}^*\mathbf{M}_{rn})}{(\nu_{pk} - \nu_0)(\nu_{rn} - \nu_0)} + \frac{(\mathbf{A}^*\mathbf{M})_{kp}\,\mathbf{M}_{pr}\,(\mathbf{A}^*\mathbf{M}_{rn})}{(\nu_{pk} - \nu_0)(\nu_{rn} + \nu_0)} \right] e^{-2\pi i \nu_{kn} t} + \tag{I, 4-23d}$$

$$+ \frac{1}{h^2} \sum_{r,p} \left[\frac{(\mathbf{A}^*\mathbf{M}_{kp})\,\mathbf{M}_{pr}\,(\mathbf{A}^*\mathbf{M}_{rn})}{(\nu_{rn} + \nu_0)(\nu_{kp} - \nu_0)} \right] e^{-2\pi i(\nu_{kn} - 2\nu_0)t} + \tag{I, 4-23e}$$

$$+ \frac{1}{h^2} \sum_{r,p} \left[\frac{(\mathbf{A}\mathbf{M}_{kp})\,(\mathbf{M}_{pr})\,(\mathbf{A}\mathbf{M}_{rn})}{(\nu_{rn} - \nu_0)(\nu_{pk} + \nu_0)} \right] e^{-2\pi i(\nu_{kn} + 2\nu_0)t} \tag{I, 4-23f}$$

and

$$\nu_{kn} = \nu_k - \nu_n\,.$$

The states p and r belong all to the orthonormal set of the unperturbed wave function and $\mathbf{M}_{nk} = \int \psi_k^* \, M \psi_n \, \mathrm{d}\tau$.

The meaning for these states is discussed later on.

Realizing that the vector **A** is complex, it is noted that the various members of the RHS of Equation (I, 4-23) are either real or complex. In order to relate these members to the classical radiation of a time-dependent real dipole moment, the relation of Klein* is employed. This relation can be formulated as follows: the radiation arising from the component

$$\mathbf{M}_{lm} e^{-2\pi i \nu_{lm} t} \tag{I, 4-24}$$

where $\mathbf{M}_{lm}$ is the moment between the ground state l and an excited state m is identical zero when $\nu_{lm} < 0$, but in the situation that $\nu_{lm} > 0$ the radiation is equal to the classically radiated intensity of the quantity

$$M_{lm} e^{-2\pi i \nu_{lm} t} + M_{lm}^* e^{2\pi i \nu_{lm} t}\,. \tag{I, 4-25}$$

Thus the first member of (I, 4-23) represents radiation ($\nu = \nu_{nk}$) of a transition $n \to k$. This transition is spontaneously induced as a result of the presence of the perturbing radiation field of frequency ν_0. Expression (I, 4-23b) represents scattered radiation of frequency $\nu_{kn} + \nu_0$, but only so if $\nu_{kn} + \nu_0 > 0$ or $W_n < W_k + h\nu_0$. The frequency difference ν_{kn} can be either positive or negative and the corresponding Raman scattered light is called Anti-Stokes and Stokes respectively. The Raman processes are depicted in Figure I-9.

Equation (I, 4-23c), on the other hand, represents scattered radiation of frequency $\nu_{kn} - \nu_0$ if $\nu_{kn} > \nu_0$ or $W_k > W_n + h\nu_0$. The latter condition can only be met if the state k is an excited state of the system. The process is depicted in Figure I-10. Emission of radiation is induced of two quanta: $h(\nu_{kn} - \nu_0)$ and $h\nu_0$ but this kind of scattering has not yet been detected, although it is possible to populate excited states to a considerable degree using optical pumping techniques with high-powered lasers as the 'pump'.

Equations (I, 4-23d), (I, 4-23e) and (I, 4-23f) represent a more complicated scattering process, which is discussed in Chapter V. Probably the most interesting scattering

* Klein, O.: 1927, *Z. Physik* **81**, 109.

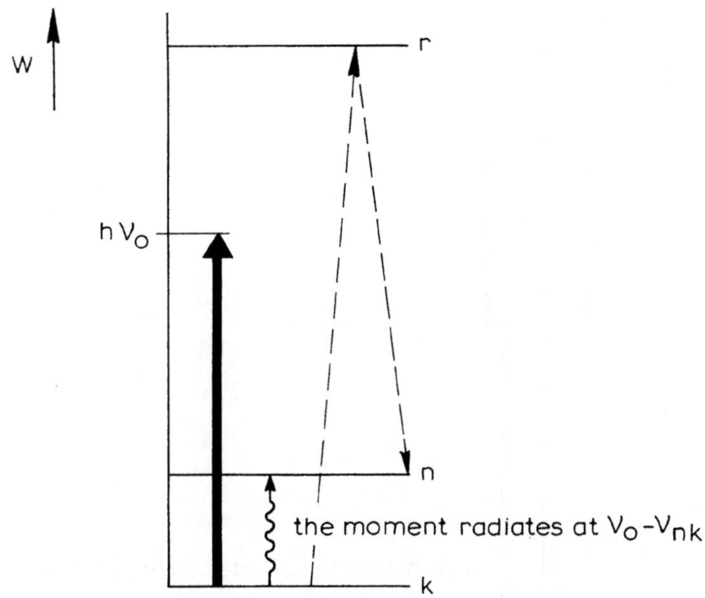

Fig. I-9a. Spontaneous Stokes Raman scattering.

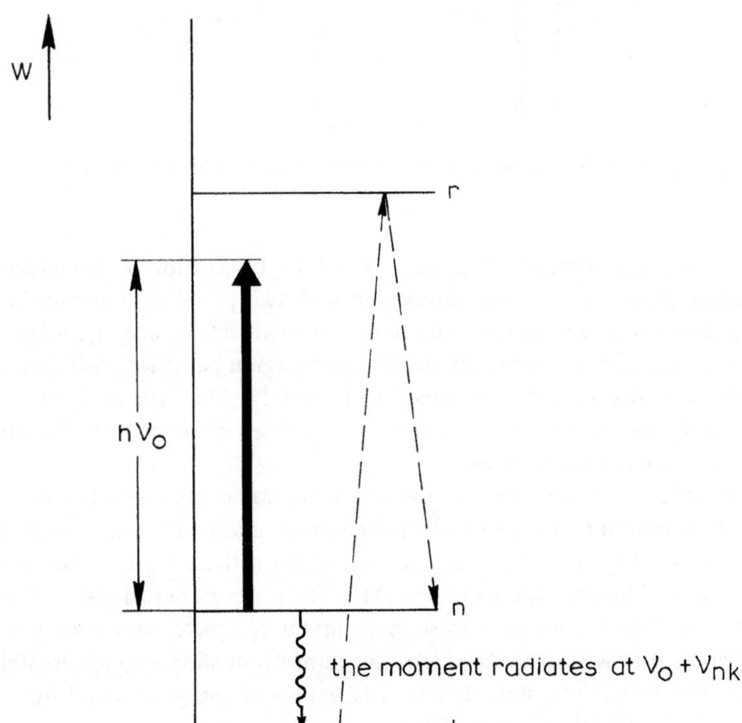

Fig. I-9b. Spontaneous anti-Stokes Raman scattering.

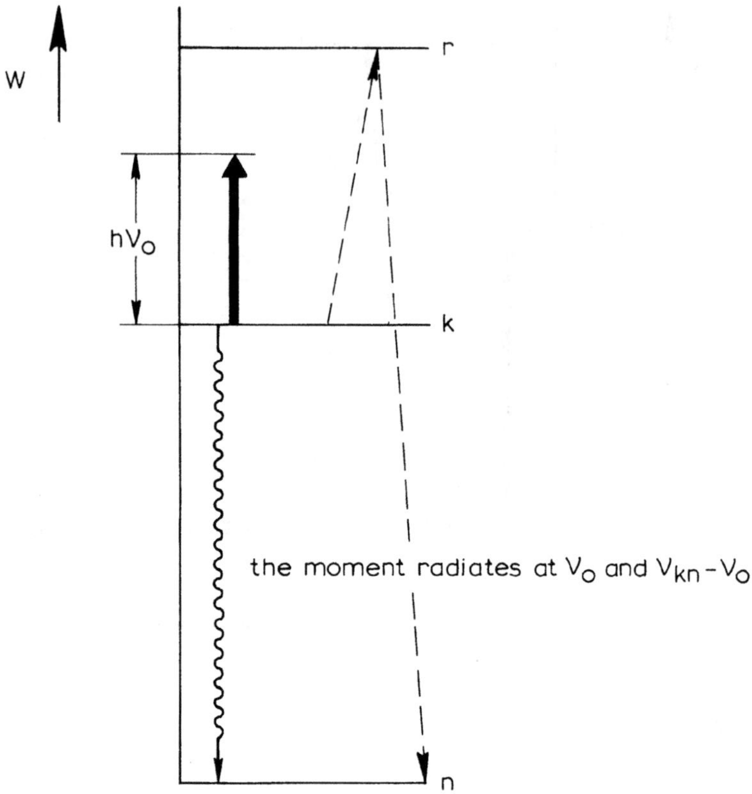

Fig. I-10. Spontaneous Raman scattering from an excited state.

process is that represented by Equation (I, 4-23f). Interaction of the molecule, atom or ion takes place not with one photon but with two photons of energy $h\nu_0$. Via the virtual states r and p, emission of radiation occurs with a frequency $\nu_{kn}+2\nu_0$ (assuming $W_n<W_k+2h\nu_0$). The frequency difference ν_{kn} can again be either positive or negative; the scattering process itself is depicted in Figure 1-11. This type of Raman scattering is called the hyper Raman effect and the effect has been observed after the introduction of lasers in Raman spectroscopy.

It is quite evident here that in the mathematical expressions for the scattering process, an important role is played by the states r and p. The summation over these states r arises solely from the fact that perturbation theory is used to derive an expression for the induced dipole moment $\mathbf{M}_{kn}$. The states r are real states of the scattering particle and the wave functions ψ_r make up the complete orthonormal set. During the scattering process the particle does not jump from state n to all possible states r and from there return to state k, nor is radiation of energy $h\nu_0$ and $h(\nu_0+\nu_{kn})$ absorbed or emitted in the classical sense.

Equation (I, 4-23b) gives only the value of a moment $\mathbf{M}_{kn}^{(1)}$. The frequency (ν) of the electromagnetic radiation emitted by this moment is that of a classical oscillating dipole moment with $\nu = \nu_0 + \nu_{kn}$ – the frequency ν_{rn} or ν_{rk} is not present here – and the direction of the moment in physical space is still not defined. It turns out that this direction is not generally parallel to the vector **E** of the incident radiation.

The vectors $\mathbf{M}_{kr}$ and $\mathbf{M}_{rn}$ represent electric dipole matrix elements of transitions between the states k and r, and r and n respectively. Only during absorption or

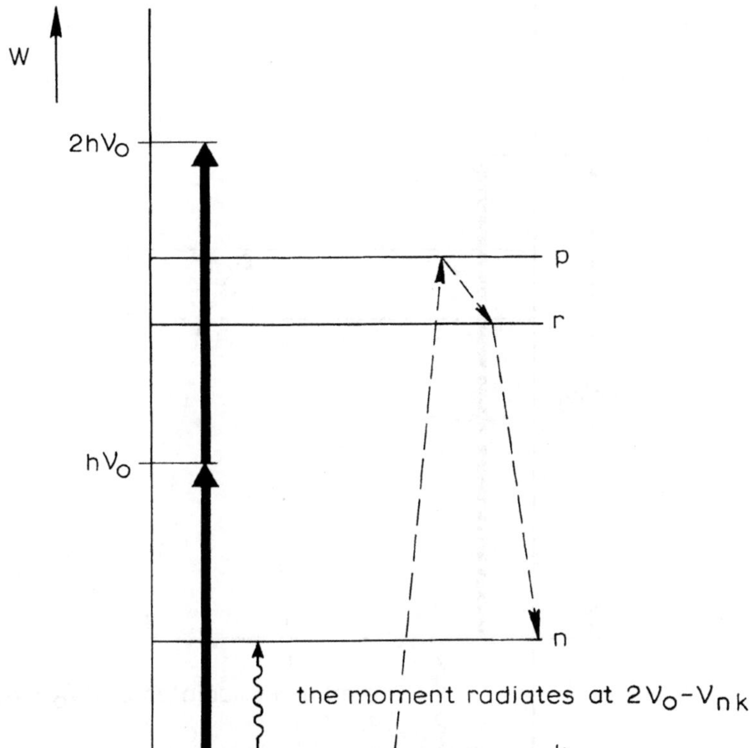

Fig. I-11a. Spontaneous Stokes hyper Raman scattering.

emission of radiation of energy $h\nu_{rk}$ and $h\nu_{nr}$ respectively is it possible to relate these matrix elements to physically observable quantities (the intensity of absorption and fluorescence bands). During the scattering process the electrons of the particle jump from the state k to an excited state n, or vice versa, as the result of a moment which is induced between these states, and the physically observable results are the Stokes and Anti-Stokes Raman bands. Equation (I, 4-23b) is related to these observable results (Stokes and Anti-Stokes Raman bands) but the intensity of these bands may be computed if values of $\mathbf{M}_{kr}$ and $\mathbf{M}_{rn}$ are inserted. This requires detailed knowledge of the complete absorption and fluorescence spectrum of the particle in the spectral region of 0Å → ∞Å. The states r are often referred to as the virtual or intermediate states, and the position of the intermediate or virtual states is immaterial. They can be either above or below the initial state k, the only condition being that $W_n < W_k + h\nu_0$. The value of the integrals $\mathbf{M}_{kr}$ or $\mathbf{M}_{rn}$ may be either negative or positive or equal to

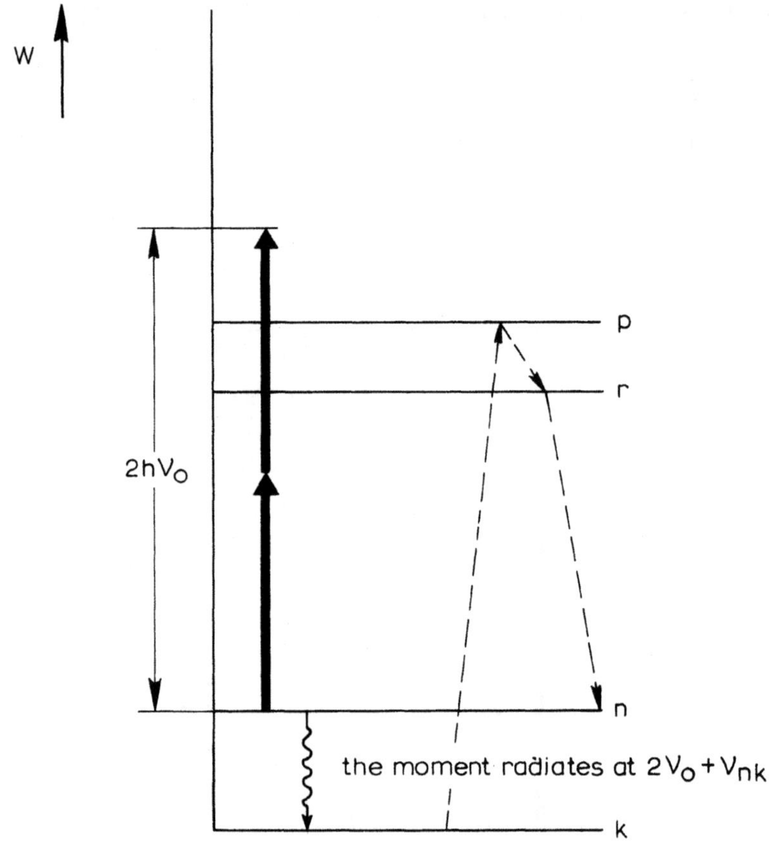

Fig. I-11b. Spontaneous anti-Stokes hyper Raman scattering.

zero. The summation over the virtual states does not always result in positive contributions.

This situation with regard to the virtual states is, of course, completely different if the energy of the incident light approaches that of an absorption band of the particle. Two things are of importance. First, the electric field now produces a large perturbation and, as a result, the Equation (I, 4-10) no longer pertains; consequently, all other expressions arising from (I, 4-10) are not valid. Secondly, the three states of the system which are now of importance are: (1) the ground state, (2) the state in which the Raman transition terminates, and (3) the excited state r which is in resonance with the energy of the incident radiation field. Any expression which attempts to describe the various resonance Raman effects – be it normal, hyper, etc. – should reflect the properties of these states and the polarization of the light (see also Chapter V).

It is evident from Equation (I, 4-23b) that the radiation emitted by the scattering molecule, atom or ion is spontaneous. A comparison with the expressions for induced (stimulated) absorption or emission of radiation (Section 3 of this chapter) proves that the aspects of induced emission are not present in mathematical form, and the conclusion is drawn that the induced moment $\mathbf{M}_{kn}$ emits spontaneous radiation of frequency $\nu_0 + \nu_{kn}$. The wave functions ψ_k and ψ_n are solutions of the time-independent Schrödinger equation (see also Equation (I, 4-16a)), and the phases of these wave functions are arbitrary. They may be designated by the factors $e^{i\delta_k}$, $e^{i\delta_n}$ and $e^{i\delta_r}$ respectively. It follows that the phases of the moments $\mathbf{M}_{kr}$ and $\mathbf{M}_{rn}$ are not necessarily equal because the phase factor of $\mathbf{M}_{kr} = \int \psi_r^* M \psi_k \, d\tau$ is equal to $e^{i(\delta_k - \delta_r)}$, and the phase factor of $\mathbf{M}_{rn}$ is then $e^{i(\delta_r - \delta_k)}$. The phase factor of the moment $\mathbf{M}_{kn}$ is then $e^{i(\delta_k - \delta_n)}$, and this factor disappears for the moment $\mathbf{M}_{kk}$. If a source emits radiation in space so that the phases of the electromagnetic waves are equal, then this radiation is called coherent radiation. A laser is the prime example of such a coherent light source. If such coherent radiation interacts with the molecule, atom or ion, then for every scattering particle in the light beam, there does not exist a difference of the phases of the induced moment $\mathbf{M}_{kk}$. The radiation of this moment is that of a classical dipole of frequency ν_0. The Rayleigh radiation Equation (I, 4-22) is thus also coherent. The particles may be seen as small coherent light sources and the waves can thus interfere with each other. The phase factor of the moment $\mathbf{M}_{kn}$ assumes different values for different molecules, atoms or ions in the light beam. If these particles are numbered 1, 2,... j, then the phase factors are $e^{i(\delta_{n_1} - \delta_{k_1})}$, $e^{i(\delta_{n_2} - \delta_{k_2})}$ ... $e^{i(\delta_{n_j} - \delta_{k_j})}$ and, in general, $\delta_{n_1} - \delta_{k_1} \neq \delta_{n_j} - \delta_{k_j}$. For the scattered radiation of frequency $\nu_0 + \nu_{kn}$, the individual particles may be considered as small incoherent light sources. The intensity of the radiation of these sources is additive, and the total intensity depends on the number of scattering centers per unit volume and is independent of the orientation of the particles. The coherence properties of Rayleigh radiation is generally destroyed as a result of density fluctuations of the scattering medium.

I-5. Formulation of the Scattering Tensor

The moment $\mathbf{M}_{kk}^{(1)}$ which is associated with the coherent Rayleigh scattering is given by:

$$\mathbf{M}_{kk}^{(1)} = \mathbf{C}_{kk} e^{-2\pi i \nu_0 t} + \mathbf{C}_{kk} e^{2\pi i \nu_0 t}. \tag{I, 5-1}$$

The total radiation emitted in space by this oscillating moment of frequency ν_0 is, according to Equation (I, 1-1):

$$I = \frac{2|\ddot{\overline{M}}|^2}{3c^3} = \frac{64\pi^4 \nu_0^4}{3c^3} |\mathbf{C}_{kk}|^2 \tag{I, 5-2}$$

where

$$\mathbf{C}_{kk} = \frac{1}{h}\sum_r \left[\frac{(\mathbf{AM}_{kr})\,\mathbf{M}_{rk}}{\nu_{rk} - \nu_0} + \frac{\mathbf{M}_{kr}\,(\mathbf{AM}_{rk})}{\nu_{rk} + \nu_0}\right]. \qquad \text{(I, 5-3)}$$

In a similar way it can be shown that the total radiation emitted by the moment $\mathbf{M}_{kn}$ which is associated with Raman scattering is equal to:

$$I = \frac{64\pi^4 (\nu_0 + \nu_{kn})^4}{3\,c^3}\,|\mathbf{C}_{kn}|^2 \qquad \text{(I, 5-4a)}$$

where

$$\mathbf{C}_{kn} = \frac{1}{h}\sum_r \left[\frac{(\mathbf{AM}_{kr})\,\mathbf{M}_{rn}}{\nu_{rk} - \nu_0} + \frac{\mathbf{M}_{kr}\,(\mathbf{AM}_{rn})}{\nu_{rn} + \nu_0}\right]. \qquad \text{(I, 5-4b)}$$

If $\mathbf{C}_{kn}$ is known, then the total intensity, cross section etc. can be computed. The moment $\mathbf{C}_{kn}$ represents a vector and if i, j and k are unit vectors along the x-, y- and z-axes of a Cartesian coordinate system, then:

$$\mathbf{C}_{kn} = (C_x)_{kn}\,\mathbf{i} + (C_y)_{kn}\,\mathbf{j} + (C_z)_{kn}\,\mathbf{k}. \qquad \text{(I, 5-5)}$$

In general, the decomposition of a vector $\mathbf{V}$ which is arbitrarily placed in space – in its components $\mathbf{v}_x$, $\mathbf{v}_y$, $\mathbf{v}_z$ along the three axes – is shown in Figure II-1. The vector $\mathbf{V}$ is here represented by a direct line segment from the origin to the point (v_x, v_y, v_z) and the magnitude of this vector is:

$$|\mathbf{v}| = (v \cdot v)^{1/2} = \sqrt{v_x^2 + v_y^2 + v_z^2}. \qquad \text{(I, 5-6)}$$

The scalar (dot or inner) product of two vectors v and u is a scalar (number) defined by:

$$\mathbf{v}\cdot\mathbf{u} = \mathbf{u}\cdot\mathbf{v} = |\mathbf{v}|\,|\mathbf{u}|\cos(\mathbf{v}, \mathbf{u}) \qquad \text{(I, 5-7)}$$

where $(\mathbf{v}, \mathbf{u})$ represents the angle between the two vectors. This product can also be written – employing the unit vectors and (I, 5-7) – as:

$$\mathbf{v}\cdot\mathbf{u} = v_x u_x + v_y u_y + v_z u_z. \qquad \text{(I, 5-8)}$$

It follows that when $\mathbf{u} \neq 0$ and $\mathbf{v} \neq 0$, and $\mathbf{v}\cdot\mathbf{u} = 0$, then u and v are perpendicular to each other.

Scalar products of vectors occur in the denominator of (I, 5-4b) because the dipole moments $\mathbf{M}_{kr}$ etc. can also be thought of as having the vector components $(M_x)_{kr}$, $(M_y)_{kr}$ and $(M_z)_{kr}$ along the Cartesian coordinates. Thus:

$$(\mathbf{AM}_{kr})\,(\mathbf{M})_{rn} = (\mathbf{AM}_{kr})\,[(M_x)_{rn}\mathbf{i} + (M_y)_{rn}\mathbf{j} + (M_z)_{rn}\mathbf{k}] \qquad \text{(I, 5-9)}$$

and

$$\mathbf{A}\cdot\mathbf{M}_{kr} = A_x\,(M_x)_{kr} + A_y\,(M_y)_{kr} + A_z\,(M_z)_{kr}. \qquad \text{(I, 5-10)}$$

For the component $(\mathbf{C}_x)_{kn}$, the following expression is obtained form (I, 5-4b) and omitting the components $\mathbf{M}_y$ and $\mathbf{M}_z$ of Equation (I, 5-9):

$$(\mathbf{C}_x)_{kn} = \frac{1}{h} \sum_r \left[\frac{(\mathbf{AM}_{kr})(\mathbf{M}_x)_{rn}}{\nu_{rk} - \nu_0} + \frac{(\mathbf{M}_x)_{kr}(\mathbf{AM}_{rn})}{\nu_{rn} + \nu_0} \right] \qquad \text{(I, 5-11)}$$

and from Equations (I, 5-9) and (I, 5-10) it follows that:

$$(\mathbf{C}_x)_{kn} = \frac{1}{h} \sum_r \left[\frac{A_x (M_x)_{kr} (M_x)_{rn} + A_y (M_y)_{kr} (M_x)_{rn} + A_z (M_z)_{kr} (M_x)_{rn}}{\nu_{rk} - \nu_0} + \right.$$
$$\left. + \frac{(M_x)_{kr} A_x (M_x)_{rn} + (M_x)_{kr} A_y (M_y)_{rn} + (M_x)_{kr} A_z (M_z)_{rn}}{\nu_{rn} + \nu_0} \right] \qquad \text{(I, 5-12)}$$

Generally speaking, by introducing ϱ and σ, where ϱ and σ stand for the Cartesian coordinates x, y or z:

$$(\mathbf{C}_\varrho)_{kn} = \frac{1}{h} \sum_r \left[\frac{\sum_\sigma \mathbf{A}_\sigma (M_\sigma)_{kr} (M_\varrho)_{rn}}{\nu_{rk} - \nu_0} + \frac{\sum_\sigma (M_\sigma)_{kr} \mathbf{A}_\sigma (M_\sigma)_{rn}}{\nu_{rn} + \nu_0} \right] \qquad \text{(I, 5-13a)}$$

so that

$$(\mathbf{C}_\varrho)_{kn} = \sum_\sigma (\alpha_{\varrho\sigma})_{kn} \mathbf{A}_\sigma \qquad \text{(I, 5-13b)}$$

where

$$(\alpha_{\varrho\sigma})_{kn} = \frac{1}{h} \sum_r \left[\frac{(M_\sigma)_{kr} (M_\varrho)_{rn}}{\nu_{rk} - \nu_0} + \frac{(M_\varrho)_{kr} (M_\sigma)_{rn}}{\nu_{rn} + \nu_0} \right]. \qquad \text{(I, 5-13c)}$$

In order to relate various induced moments to the direction of propagation and polarization of the incident light, it is helpful to return to Equation (I, 4-9). Thus, for linear-polarized radiation, with the electric vector $\mathbf{E}_0$ parallel to the z-axis, and its direction of propagation along the x-axis, it is found that:

$$A_z = \tfrac{1}{2} E_0 e^{-i\varphi} \qquad \text{(I, 5-14a)}$$

and

$$I_0 = \frac{c}{8\pi} E_0^2; \qquad |A_z|^2 = \frac{E_0^2}{4} \qquad \text{(I, 5-14b)}$$

The maximum value of A_z is $\tfrac{1}{2}E_0$. From Equation (I, 5-13c) it follows that:

$$(\alpha_{xz})_{kn} = \frac{1}{h} \sum_r \left[\frac{(M_z)_{kr} (M_x)_{rn}}{\nu_{rk} - \nu_0} + \frac{(M_x)_{kr} (M_z)_{rn}}{\nu_{rn} + \nu_0} \right] \qquad \text{(I, 5-15)}$$

and according to Equation (I, 5-4a):

$$I_{kn} = \frac{128\,\pi^5 (\nu_0 + \nu_{kn})^4}{3c^4} (\alpha_{xz})_{kn}^2 I_0 . \qquad \text{(I, 5-16)}$$

It is also possible to find expression for the moments in the y and z directions employing (I, 5-13b). The result is that:

$$(\mathbf{C}_y)_{kn} = \alpha_{yz}\mathbf{A}_z \quad \text{and} \quad (\mathbf{C}_z)_{kn} = \alpha_{zz}\mathbf{A}_z\,. \tag{I, 5-17}$$

Finally, the following additional expressions are obtained in evaluating the moments in x, y and z directions induced by incident radiation polarized along the z and y axes:

$$\begin{aligned} (\mathbf{C}_x)_{kn} &= \alpha_{xy}\mathbf{A}_y & \qquad (\mathbf{C}_x)_{kn} &= \alpha_{xx}\mathbf{A}_x \\ (\mathbf{C}_y)_{kn} &= \alpha_{yy}\mathbf{A}_y \quad \text{and} & (\mathbf{C}_y)_{kn} &= \alpha_{yx}\mathbf{A}_x \\ (\mathbf{C}_z)_{kn} &= \alpha_{zy}\mathbf{A}_y & (\mathbf{C}_z)_{kn} &= \alpha_{zx}\mathbf{A}_x\,. \end{aligned} \tag{I, 5-18}$$

The Equation (I, 5-13a) combines the three different directions of polarization of the incident light with induced dipole moments again along three different directions – the three Cartesian axes. It is seen that, in total, nine relations reflect the different ways in which combinations can be made; and the nine elements $\alpha_{\varrho\sigma}$ are the elements of a tensor of order two. This tensor is normally written as:

$$\begin{pmatrix} \alpha_{xx} & \alpha_{xy} & \alpha_{xz} \\ \alpha_{yx} & \alpha_{yy} & \alpha_{yz} \\ \alpha_{zx} & \alpha_{zy} & \alpha_{zz} \end{pmatrix} \tag{I, 5-19}$$

and the following are the relations of the moments and the components of the field:

$$\begin{aligned} (\mathbf{C}_x)_{kn} &= (\alpha_{xx})_{kn}\mathbf{A}_x + (\alpha_{xy})_{kn}\mathbf{A}_y + (\alpha_{xz})_{kn}\mathbf{A}_z \\ (\mathbf{C}_y)_{kn} &= (\alpha_{yx})_{kn}\mathbf{A}_x + (\alpha_{yy})_{kn}\mathbf{A}_y + (\alpha_{yz})_{kn}\mathbf{A}_z \\ (\mathbf{C}_z)_{kn} &= (\alpha_{zx})_{kn}\mathbf{A}_x + (\alpha_{zy})_{kn}\mathbf{A}_y + (\alpha_{zz})_{kn}\mathbf{A}_z\,. \end{aligned} \tag{I, 5-20}$$

The tensor α determines the spatial orientation of the moments $(\mathbf{C}_\varrho)_{kn}$. If all tensor elements are different from zero, then it can be seen that incident radiation, propagating parallel to the x-axis of a space fixed coordinate system, with the electric vector parallel to the z-axis, induces moments in the x, y and z direction. The magnitude of these moments is determined by the components α_{xz}, α_{yz} and α_{zz} respectively. The orientation of the moment $\mathbf{C}_{kn}$ is thus not always parallel to the z-axis, as is indicated in Figure I-12. The total intensity I_{kn} radiated in space is, according to Equations (I, 5-4), (I, 5-13) and (I, 5-14):

$$I_{kn} = \frac{128\,\pi^5\,(\nu_0 + \nu_{kn})^4}{3c^4} \left| \sum_\sigma (\alpha_{\varrho\sigma})_{kn} \right|^2 I_0 \tag{I, 5-21}$$

which can also be written as:

$$I_{kn} = Q_{kn} I_0 \tag{I, 5-22a}$$

where

$$Q_{kn} = \frac{128\,\pi^5\,(\nu_0 + \nu_{kn})^4}{3c^4} \left| \sum_\sigma (\alpha_{\varrho\sigma})_{kn} \right|^2 . \tag{I, 5-22b}$$

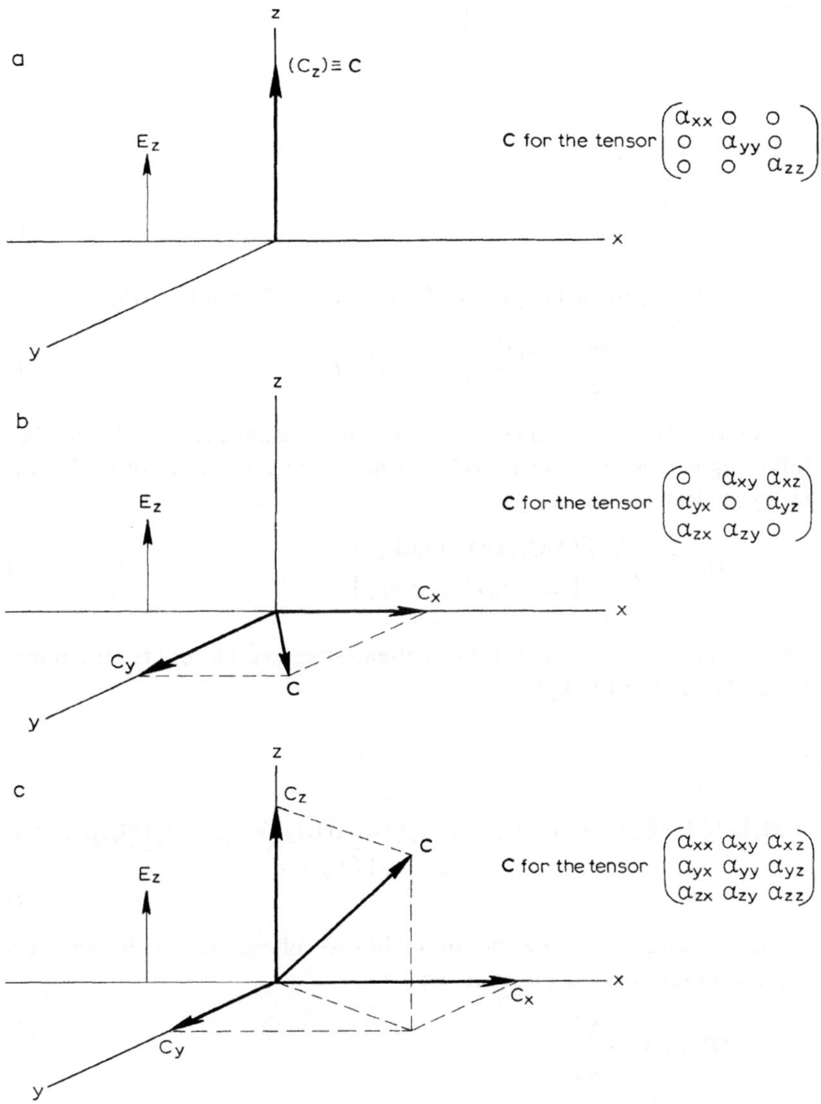

Fig. I-12. The direction of the moment **C** in space, for different tensors, if the incident light is polarized parallel to the z-axis ($E_z = A_z \neq 0$).

The quantity Q_{kn} has the dimension of an area and is called the cross section of the scattered radiation. Q depends on the polarization of the incident radiation and Equation (I, 5-22) reflects the situation where this radiation is plane-polarized.

It is sometimes more convenient to have at hand an expression of the intensity emitted per unit solid angle in a direction perpendicular to the moment $\mathbf{C}_{kn}$. According to Equations (I, 1-13) and (I, 1-1), the intensity per unit time in the solid angle dΩ is:

$$\mathrm{d}I = \frac{\overline{\ddot{M}}^2}{4\,\pi c^3} \sin^2\theta \,\mathrm{d}\Omega \tag{I, 5-23a}$$

and per unit solid angle:

$$I = \frac{\overline{\ddot{M}}^2}{4\,\pi c^3} \sin^2\theta \tag{I, 5-23b}$$

and thus for $\theta = \frac{1}{2}\pi$, by making use of Equation (I, 5-2) and (I, 5-16)

$$I_{kn} = \frac{16\,\pi^4 (\nu_0 + \nu_{kn})^4}{3c^4} \left| \sum_\sigma (\alpha_{\varrho\sigma})_{kn} \right|^2 I_0\,. \tag{I, 5-24}$$

The tensor for the hyper Raman effect is more complicated. Similar to the above, the following expression – see (I, 4-23f) – has to be considered for Stokes and anti-Stokes hyper Raman scattering:

$$\mathbf{D}_{kn} = \frac{1}{h^2} \sum_{r,\,p} \left[\frac{(\mathbf{AM}_{kp})\,\mathbf{M}_{pr}\,(\mathbf{AM}_{rn})}{(\nu_{rn} - \nu_0)\,(\nu_{pk} + \nu_0)} \right]. \tag{I, 5-25}$$

For the moment $(\mathbf{D}_x)_{kn}$ can then be written, after evaluating first the inner vector products $(\mathbf{AM}_{nn})$ and $(\mathbf{AM}_{np})$:

$$(\mathbf{D}_x)_{kn} = \frac{1}{h^2} \sum_{r,\,p}$$
$$\frac{(\mathbf{M}_x)_{pr} [(A_x (M_x)_{kp} + A_y (M_y)_{kp} + A_z (M_z)_{kp})\,(A_x (M_x)_{pr} + A_y (M_y)_{pr} + A_z (M_z)_{pr})]}{(\nu_{rn} - \nu_0)\,(\nu_{pk} + \nu_0)} \tag{I, 5-26}$$

Generally speaking, following the above in case of the normal Raman scattering, the expression for all three moments is:

$$(\mathbf{D}_\sigma)_{kn} = \frac{1}{h^2} \sum_{r,\,p}$$
$$(M_\sigma)_{pr} \left[\sum_\sigma (A_\sigma A_\sigma (M_\sigma)_{pr} (M_\sigma)_{rn}) + \sum_{\sigma \neq \varrho} A_\sigma A_\varrho (M_\sigma)_{rn} (M_\varrho)_{pr}\right] \tag{I, 5-27}$$

As a result of the fact that two incident photons combine, the following expressions are obtained for the moments:

$$\begin{aligned}
(\mathbf{D}_x)_{kn} &= \beta_{xxx}\mathbf{A}_x\mathbf{A}_x + \beta_{xxy}\mathbf{A}_x\mathbf{A}_y + \beta_{xxz}\mathbf{A}_x\mathbf{A}_z + \beta_{xyx}\mathbf{A}_y\mathbf{A}_x + \cdots \beta_{xzz}\mathbf{A}_z\mathbf{A}_z \\
(\mathbf{D}_y)_{kn} &= \beta_{yxx}\mathbf{A}_x\mathbf{A}_x + \beta_{yxy}\mathbf{A}_x\mathbf{A}_y + \beta_{yxz}\mathbf{A}_x\mathbf{A}_z + \beta_{yyx}\mathbf{A}_y\mathbf{A}_x + \cdots \beta_{yzz}\mathbf{A}_z\mathbf{A}_z \\
(\mathbf{D}_z)_{kn} &= \beta_{zxx}\mathbf{A}_x\mathbf{A}_x + \beta_{zxy}\mathbf{A}_x\mathbf{A}_y + \beta_{zxz}\mathbf{A}_z\mathbf{A}_z + \beta_{zyx}\mathbf{A}_y\mathbf{A}_x + \cdots \beta_{zzz}\mathbf{A}_z\mathbf{A}_z
\end{aligned} \tag{I, 5-28}$$

The elements $\beta_{\varrho\sigma\sigma'}$ are the elements of a tensor of order three, and part of the general expression is:

$$\beta_{\varrho\sigma\sigma'} = \frac{1}{h^2} \sum_{r,\,p} \frac{(M_{\sigma'})_{rk} (M_\sigma)_{pr} (M_\varrho)_{kp}}{(\nu_{rk} + \nu_0)(\nu_{pn} - \nu_0)}. \tag{I, 5-29}$$

I-6. Symmetry of the Scattering Tensors

A detailed discussion of properties of tensors is given in Chapter II, but of relevance here is the symmetry of the tensor of the various types of scattered light; in addition, the question has to be answered whether these tensors are real or complex. In discussing these points, it seems convenient to return to the expressions derived for the coherent Rayleigh radiation. The scattered light is related to the moment $\mathbf{M}_{kk}$ where

$$\mathbf{M}_{kk}^{(1)} = \mathbf{C}_{kk} e^{-2\pi i \nu_0 t} + \mathbf{C}_{kk}^* e^{2\pi i \nu_0 t} \tag{I, 6-1a}$$

and

$$\mathbf{C}_{kk} = \frac{1}{h} \sum_r \left[\frac{(\mathbf{AM}_{kr})(\mathbf{M}_{rk})}{\nu_{rk} - \nu_0} + \frac{\mathbf{M}_{kr}(\mathbf{AM}_{rk})}{\nu_{rk} + \nu_0} \right]. \tag{I, 6-1b}$$

Similar to the case of the Raman scattered light and after finding the expressions for the moment C_{kk} in x, y and z directions, these moments may again be related to the various components of polarization of the incident light. The results are that:

$$(\mathbf{C}_\varrho)_{kk} = \sum_\sigma (\alpha_{\varrho\sigma})_{kk} \, \mathbf{A}_\sigma \tag{I, 6-1c}$$

where

$$(\alpha_{\varrho\sigma})_{kk} = \frac{1}{h} \sum_r \left[\frac{(M_\sigma)_{kr} (M_\varrho)_{rk}}{\nu_{rk} - \nu_0} + \frac{[\varrho \leftrightarrow \sigma]}{\nu_{rk} + \nu_0} \right]. \tag{I, 6-1d}$$

Although the expressions for $(\alpha_{\varrho\sigma})_{kk}$ may be obtained by following the method outlined in the previous paragraph, an extremely simple way to arrive at (I, 6-1d) is to let $n=k$ in (I, 5-13b).

The integral $\int \psi_k^* M \psi_r \, d\tau$ is called Hermitian because

$$\left[\int \psi_k^* M \psi_r \, d\tau \right]^* = \int \psi_r^* M \psi_k \, d\tau$$

or

$$[M_{rk}]^* = M_{kr}. \tag{I, 6-2}$$

After substituting (1, 6-2) in (I, 6-1d) the following expression is obtained for the complex conjugate elements $(\alpha_{\varrho\sigma})_{kk}^*$:

$$(\alpha_{\varrho\sigma})_{kk}^* = \frac{1}{h} \sum_r \left[\frac{(M_\sigma)_{kr} (M_\varrho)_{rk}}{\nu_{rk} - \nu_0} + \frac{[\varrho \leftrightarrow \sigma]}{\nu_{rk} + \nu_0} \right]. \tag{I, 6-3}$$

From (I, 6-1d) and (I, 6-3) it follows that:

$$(\alpha_{\varrho\sigma})_{kk} = (\alpha_{\varrho\sigma})^*_{kk}. \qquad \text{(I, 6-4)}$$

The elements are also Hermitian. An important relation exists if the wave functions of the states k and r of the scattering particle are all real. From (I, 6-2) it is seen that $[M_{kr}]^* = M_{kr}$ and consequently $(\alpha_{\varrho\sigma})_{kk} = (\alpha_{\sigma\varrho})_{kk}$. The tensor under these conditions is symmetric; that is:

$$\begin{aligned} &\alpha_{xy} = \alpha_{yx}; \quad \alpha_{xz} = \alpha_{zx}; \quad \alpha_{yz} = \alpha_{zy} \\ &\alpha_{xx} \neq \alpha_{yy} \neq \alpha_{zz} \end{aligned} \qquad \text{(I, 6-5)}$$

but, in general, the opposite is also true; if the tensor is symmetric, then the elements are all real. There does exist another situation where real tensors are encountered. For the theoretical case where $\nu_0 = 0$ – the so-called Static Field – the numerators of both members of (I, 6-1d) become identical and here again it is seen that:

$$(\alpha_{\varrho\sigma})_{kk} = (\alpha_{\sigma\varrho})_{kk}. \qquad \text{(I, 6-6)}$$

For real values of the elements of the scattering tensor for $\nu_0 \neq 0$ and by combining (I, 6-1a) and (I, 6-1c), it follows that:

$$\mathbf{M}^{(1)}_{kk} = \sum_\sigma (\alpha_{\varrho\sigma})_{kk} \left(\mathbf{A}_\sigma e^{-2\pi i \nu_0 t} + \mathbf{A}^*_\sigma e^{2\pi i \nu_0 t}\right) \qquad \text{(I, 6-7)}$$

and, employing the expressions (I, 4-6) and (I, 4-7), it follows that:

$$\mathbf{M}^{(1)}_{kk} = \sum_\sigma (\alpha_{\varrho\sigma})_{kk} \mathbf{E}_\sigma. \qquad \text{(I, 6-7a)}$$

The real elements $(\alpha_{\varrho\sigma})_{kk}$ are called the elements of the polarizability tensor α where α_{kk} is the polarizability of the ground state k. The polarizability tensor is a special case of the more general scattering tensor which is, in principle, asymmetric even if the wave functions are real.

For the Raman effect where interaction occurs with one photon of energy (see (I, 5-23b), (I, 5-23c)), the elements of the scattering tensor are:

$$(\alpha_{\varrho\sigma})_{kn} = \frac{1}{h} \sum_r \left[\frac{(M_\sigma)_{kr}(M_\varrho)_{rn}}{\nu_0 - \nu_{rk}} + \frac{[\varrho \leftrightarrow \sigma]}{\nu_0 + \nu_{rn}}\right] \qquad \text{(I, 6-8a)}$$

and

$$\mathbf{M}^{(1)}_{kn} = \sum_\sigma (\alpha_{\varrho\sigma})_{kn} \mathbf{E}_\sigma. \qquad \text{(I, 6-8b)}$$

The Equation (I, 6-8b) includes the Stokes and anti-Stokes Raman effect and the excited state Raman process (see also Figures I-9 and I-10).

For the Raman effect where interaction takes place between the molecule and two photons of energy $h\nu_0$:

$$(\beta_{\varrho\sigma\sigma'})_{kn} = \frac{1}{h^2} \sum_{rp} \frac{(M_{\sigma'})_{rk}(M_\sigma)_{pr}(M_p)_{kp}}{(\nu_{rn} - \nu_0)(\nu_{pk} + \nu_o)} \qquad \text{(I, 6-9a)}$$

and

$$\mathbf{M}_{kn}^{(2)} = \sum_{\varrho\sigma\sigma'} \beta_{\varrho\sigma\sigma'} \mathbf{E}_{\sigma} \mathbf{E}_{\sigma}.$$

It is noted that even in the situation that all wave functions are real, so that $(M_\sigma)_{rk} = (M_\sigma)_{kr}$ etc., the occurrence of the equality $(M_\sigma)_{kr} = (M_\varrho)_{kr}$ is accidental, so that $(\alpha_{\varrho\sigma})_{kn} \neq (\alpha_{\sigma\varrho})_{kn}$ and $(\beta_{\varrho\sigma\sigma'})_{kn} \neq (\beta_{\sigma\varrho\sigma'})_{kn} \neq (\beta_{\sigma\sigma\varrho})_{kn}$; and the conclusion is reached that both tensors are not symmetric. Real eigenfunctions ψ_k and ψ_r are obtained as solutions of a Schrödinger equation if the Hamiltonian $\mathscr{H}$ is real.

The scattering tensor α for the normal Raman effect has nine entries which are not equal to each other if the tensor is asymmetric. The symmetric tensor has six independent entries. For the hyper Raman effect there are 27 entries. If β is asymmetric, then there are 18 independent entries. This number is reduced to 10 if the hyper Raman tensor is symmetric.

Suggestions for Further Reading

Born, M., Heisenberg, W., and Jordan, P. J.: 1925, *Z. Physik* **35**, 557.
Cabannes, J.: 1928, *Compt. Rend.* **186**, 1201.
Cabannes, J. and Rocard, Y.: 1929, *J. Phys. Radium* **10**, 52.
Dirac, P. A. M.: 1930, *Quantum Mechanics*, Oxford.
Heisenberg, W.: 1931, *Ann. Physik*, **9**, 338.
Kramers, H. A. and Heisenberg, W.: 1925, *Z. Phys.* **31**, 681.
Landsberg, G. and Mandelshtam, L.: 1928, *Naturwiss.* **16**, 557.
Lommel, E.: 1871, *Poggendorffs Ann.* **143**, 26.
Manneback, C.: 1930, *Z. Physik* **62**, 224.
Placzek, G.: 1931, *Z. Physik* **70**, 84.
Placzek, G.: 1931, *Leibziger Vortrage*, 71.
Placzek, G.: 1934, *Marx Handbuch der Radiologie* (2d ed.), vol. VI, part II, pp. 209–374.
Placzek, G. and Teller, E.: 1933, *Z. Physik* **83**, 209.
Raman, C. V.: 1928, *Indian J. Phys.* **2**, 387.
Raman, C. V. and Krishnan, R. S.: 1928, *Nature* **121**, 501.
Rasetti, F.: 1931, *Leibziger Vortrage*, 59.
Schrödinger, E.: 1926, *Ann. Physik* **81**, 109.
Smekal, A.: 1923, *Naturwiss.* **11**, 873.
Van Vleck, J. H.: 1929, *Proc. Nat. Acad. Sci.* **15**, 754.
Wood, W. R.: 1908, *Phys. Z.* **9**, 450.

CHAPTER II

PROPERTIES OF TENSORS

In the preceding chapter was indicated that the quantity which relates the induced moments to the components of the incident electric field is called a tensor. This tensor lies at the heart of Rayleigh and Raman scattering, and it appears that a knowledge of the properties of tensors is of importance in describing the properties of the scattered light.

The scattering tensor is not unique. In engineering and physics occur other tensors like the stress or strain tensor, the moment of the inertia tensor, and in atomic physics, the g tensor. The former arises in connection with the study of the application of an external force to a body. The displacement of this body – under influence of the force – does not need to be parallel to the direction of the applied force, and this effect is traced back to resistive forces which are anisotropic. Tensors, or dyadics as they are officially called, can be very simple; and the easiest tensor is a scalar. A scalar is said to be a tensor of rank (or order) zero. Vectors are also examples of tensors. A normal vector is a tensor of rank one. The scattering tensor and stress tensor are examples of tensors of rank two. Such tensors are also called dyadics. A polyadic is a tensor of higher order, it is encountered in describing the hyper Raman effect, but the basis of all tensors is associated with the algebra of vectors and matrices.

II-1. Vectors and Dyadics

It may be recalled here that in the elementary vector analysis two different products are defined. There is a scalar or dot product given by

$$\mathbf{u} \cdot \mathbf{v} \tag{II, 1-1}$$

and the second product is called the cross product

$$\mathbf{u} \times \mathbf{v} . \tag{II, 1-2}$$

The scalar product of the vectors $\mathbf{u}$ and $\mathbf{v}$ is defined as

$$\mathbf{u} \cdot \mathbf{v} = |\mathbf{u}| \, |\mathbf{v}| \cos(\mathbf{u}, \mathbf{v}) \tag{II, 1-3}$$

where $(\mathbf{u}, \mathbf{v})$ is the angle between the two vectors and the magnitude $|v|$ is given by:

$$|\mathbf{v}| = (v \cdot v)^{1/2} = (v_x^2 + v_y^2 + v_z^2)^{1/2} . \tag{II, 1-4}$$

The vector $\mathbf{v}$ in physical space is represented by a direct line segment from the origin of a Cartesian coordinate system to the point (v_x, v_y, v_z). If $\mathbf{i}$, $\mathbf{j}$ and $\mathbf{k}$ are unit vectors in the x, y and z directions (see Figure II-1) and v_x, v_y and v_z are the components, then:

$$\mathbf{v} = \mathbf{i}v_x + \mathbf{j}v_y + \mathbf{k}v_z\,. \tag{II, 1-5}$$

The dot product can also be written as:

$$\mathbf{u}\cdot\mathbf{v} = u_x v_x + u_y v_y + u_z v_z\,. \tag{II, 1-6}$$

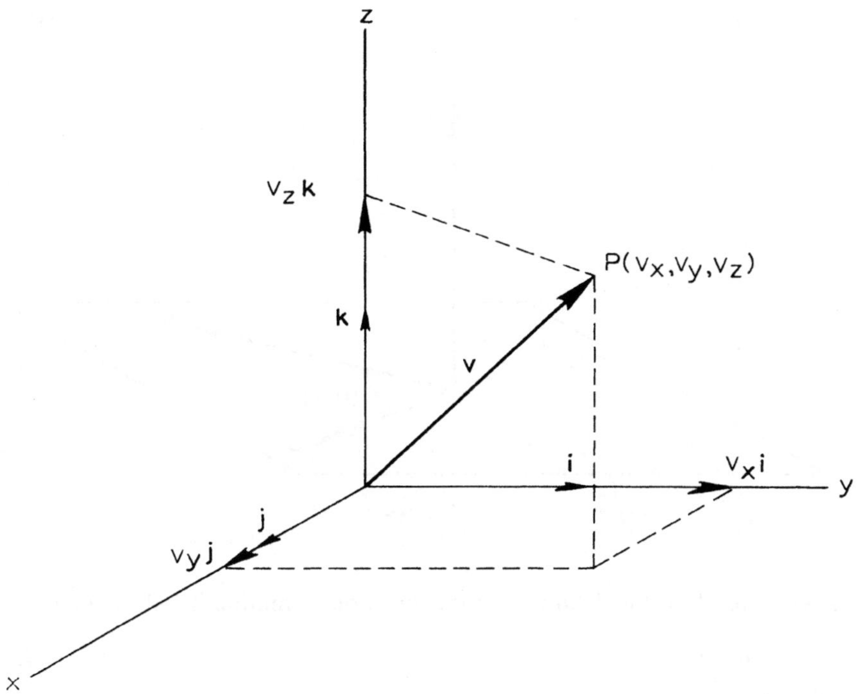

Fig. II-1.

In the situation that the vectors are perpendicular (orthogonal) to each other, it follows from (II, 1-3) that

$$\mathbf{u}\cdot\mathbf{v} = 0 \tag{II, 1-7}$$

and the following relations hold for the unit vectors **i**, **j** and **k**;

$$\begin{aligned} \mathbf{i}\cdot\mathbf{i} &= 1 \qquad & \mathbf{i}\cdot\mathbf{j} &= \mathbf{j}\cdot\mathbf{i} = 0 \\ \mathbf{j}\cdot\mathbf{j} &= 1 \qquad & \mathbf{j}\cdot\mathbf{k} &= \mathbf{k}\cdot\mathbf{j} = 0 \\ \mathbf{k}\cdot\mathbf{k} &= 1 \qquad & \mathbf{i}\cdot\mathbf{k} &= \mathbf{k}\cdot\mathbf{i} = 0 \end{aligned} \tag{II, 1-8}$$

The distributive law holds for the dot product, thus:

$$\mathbf{u}\cdot(a\mathbf{v} + b\mathbf{w}) = a\mathbf{u}\cdot\mathbf{v} + b\mathbf{u}\cdot\mathbf{w}\,. \tag{II, 1-9}$$

On the other hand, the vector or cross product of **u** and **v** is defined as:

$$\mathbf{u} \times \mathbf{v}. \qquad \text{(II, 1-10)}$$

This product is defined in terms of a new vector, for example **w**, which is perpendicular to **u** and **v** and the direction of motion of a right-handed screw when turned from **u** to **v** (see Figure II-2). The magnitude of **w** is defined by:

$$|\mathbf{w}| = |\mathbf{u}|\,|\mathbf{v}| \sin(\mathbf{u}, \mathbf{v}). \qquad \text{(II, 1-11)}$$

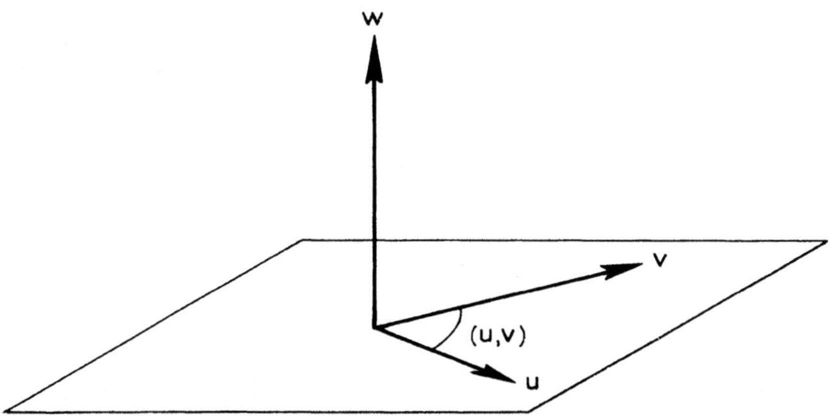

Fig. II-2.

As a direct result of the definition of the direction of motion it follows that:

$$\mathbf{u} \times \mathbf{v} = -\,\mathbf{v} \times \mathbf{u} \qquad \text{(II, 1-12)}$$

but also

$$\mathbf{u} \times \mathbf{u} = \mathbf{v} \times \mathbf{v} = 0. \qquad \text{(II, 1-12a)}$$

The unit vectors **i**, **j** and **k**, directed towards the x-, y- and z-axes, may also be used to form cross products. Thus:

$$\mathbf{i} \times \mathbf{i} = \mathbf{j} \times \mathbf{j} = \mathbf{k} \times \mathbf{k} = 0 \qquad \text{(II, 1-13)}$$

$$\begin{aligned} \mathbf{i} \times \mathbf{j} &= \mathbf{k} = -\,\mathbf{j} \times \mathbf{i} \\ \mathbf{k} \times \mathbf{i} &= \mathbf{j} = -\,\mathbf{i} \times \mathbf{k} \\ \mathbf{j} \times \mathbf{k} &= \mathbf{i} = -\,\mathbf{k} \times \mathbf{j} \end{aligned} \qquad \text{(II, 1-13a)}$$

Employing these unit vectors, the vector product of **u** and **v** can also be formulated as:

$$\begin{aligned} \mathbf{u} \times \mathbf{v} &= (\mathbf{i}u_x + \mathbf{j}u_y + \mathbf{k}u_z) \times (\mathbf{i}v_x + \mathbf{j}v_y + \mathbf{k}v_z) \\ &= \mathbf{i}(u_y v_z - u_z v_y) + \mathbf{j}(u_z v_x - u_x v_z) + \mathbf{k}(u_x v_y - u_y v_x) \end{aligned} \qquad \text{(II, 1-14)}$$

and this particular cross product in determinantal form is:

$$\mathbf{u} \times \mathbf{v} = \begin{vmatrix} \mathbf{i} & \mathbf{j} & \mathbf{k} \\ u_x & u_y & u_z \\ v_x & v_y & v_z \end{vmatrix} \tag{II, 1-15}$$

A rectangular array of m rows and n columns of numbers or other quantities is called a matrix. Thus a matrix A may be written as:

$$A = \begin{pmatrix} a_{11}a_{12}a_{13} \\ a_{21}a_{22}a_{23} \end{pmatrix} \tag{II, 1-16}$$

but a determinant is a quantity associated only with a *square array* of n^2 elements. Thus:

$$|A| = \begin{vmatrix} a_{11} & a_{12} \\ a_{21} & a_{22} \end{vmatrix}$$

is a determinant and the quantity is equal to $a_{11}a_{22} - a_{12}a_{21}$. Equation (II, 1-14) represents the value of the determinant given by II, 1-15.

Matrices may also be multiplied together. The product C of two matrices A and B is defined as follows:

$$C = AB \tag{II, 1-16a}$$

and the elements of the matrix C are obtained in the following way:

$$C_{ij} = \sum_k A_{ik} B_{kj}\,. \tag{II, 1-16b}$$

The subscript i refers to the row, and j to the column. It does not necessarily follow that $AB = BA$, and it is said that A does not commute with B if $AB \neq BA$. A diagonal matrix has only elements on the principal diagonal (running from the upper left to the lower right corner); a special diagonal matrix is the unit matrix E. The diagonal elements of this matrix are all equal to unity and thus:

$$EA = AE = A\,. \tag{II, 1-17}$$

The inverse of reciprocal of the matrix A is denoted by A^{-1} and defined in the following way:

$$A^{-1} \cdot A = A \cdot A^{-1} = E\,. \tag{II, 1-17a}$$

However not all matrices have the property so that Equation (II, 1-17a) is satisfied.

An important property of square matrices, which follows directly from laws governing the addition and subtraction of such matrices, is the following: Any square matrix may be given as the sum of a symmetrical (A) and antisymmetrical (A^a) matrix. Thus if

$$A = \begin{pmatrix} a_{11} & a_{12} \\ a_{21} & a_{22} \end{pmatrix} \quad \text{and} \quad A^a = \begin{pmatrix} a_{11} & a_{21} \\ a_{12} & a_{22} \end{pmatrix} \tag{II, 1-18a}$$

then

$$A = \frac{A + A^a}{2} + \frac{A - A^a}{2}. \qquad \text{(II, 1-18b)}$$

The elements of the matrices are:

$$\frac{A + A^a}{2} = \begin{pmatrix} a_{11} & (a_{12} + a_{21})/2 \\ (a_{21} + a_{12})/2 & a_{22} \end{pmatrix} \qquad \text{(II, 1-18c)}$$

and

$$\frac{A - A^a}{2} = \begin{pmatrix} a_{11} & (a_{12} - a_{21})/2 \\ (a_{21} - a_{12})/2 & a_{22} \end{pmatrix} \qquad \text{(II, 1-18d)}$$

A^a is the transpose A^T of the matrix A and is defined by having the elements of A by interchanging all rows with corresponding columns. There also exists a complex conjugate of A. This matrix is denoted by A^* and the elements of this matrix are equal to the complex conjugate of the corresponding elements of A. In addition, there may also be defined the conjugate transpose matrix $A^\dagger$ so that

$$A^\dagger = A^{T*} \quad \text{or} \quad A^\dagger_{ij} = A^{T*}_{ij} = A^*_{ji}.$$

A matrix is called Hermetian if $A = A^T$. If this condition is met and at the same time all elements of A are real, then the matrix A must be symmetric around the principal diagonal. Furthermore, the matrix is said to be orthogonal if $A^{-1} = A^T$ so that $AA^T = E$ also. For the elements, the following relation holds: $(A^{-1})_{ji} = A_{ij}$. It may also be possible that the condition is met: $(A^{-1})^*_{ji} = A_{ij}$. If this happens, then the matrix A is called unitary and thus $A^{-1} = A^\dagger$ so that $AA^\dagger = E$. It may also be seen that a real unitary matrix is equivalent to an orthogonal matrix.

Square matrices like (II, 1-18a), have characteristic values or eigenvalues. They are obtained by allowing the determinant to be equal to zero. Thus, for a matrix of three rows and columns, the eigenvalues are obtained from:

$$\begin{vmatrix} a_{11} - \lambda & a_{12} & a_{13} \\ a_{21} & a_{22} - \lambda & a_{23} \\ a_{33} & a_{32} & a_{33} - \lambda \end{vmatrix} = 0. \qquad \text{(II, 1-19a)}$$

Such a determinantal equation may occur if a solution is sought for an equation of the third degree in the unknown λ, but arises also as a condition that the three simultaneous equations involving v_1, v_2 and v_3 as unknowns:

$$\begin{aligned} a_{11}v_1 + a_{12}v_2 + a_{13}v_3 &= \lambda_1 v_1 \\ a_{21}v_1 + a_{22}v_2 + a_{23}v_3 &= \lambda_2 v_2 \\ a_{31}v_1 + a_{23}v_2 + a_{33}v_3 &= \lambda_2 v_3 \end{aligned} \qquad \text{(II, 1-19b)}$$

should possess a solution other than the trivial one ($v_1 = v_2 = v_3 = 0$). (II, 1-19b) may also be written in matrix notation as:

$$\begin{pmatrix} a_{11} & a_{12} & a_{13} \\ a_{21} & a_{22} & a_{23} \\ a_{31} & a_{32} & a_{33} \end{pmatrix} \begin{pmatrix} v_1 \\ v_2 \\ v_3 \end{pmatrix} = \begin{pmatrix} \lambda_1 & & \\ & \lambda_2 & \\ & & \lambda_3 \end{pmatrix} \begin{pmatrix} v_1 \\ v_2 \\ v_3 \end{pmatrix}. \tag{II, 1-19c}$$

Vectors in three-dimensional space may also be represented by matrices. The vector **u** defined as

$$\mathbf{u} = \mathbf{i}u_x + \mathbf{j}u_y + \mathbf{k}u_z \tag{II, 1-20a}$$

is represented by the following row or column matrices:

$$U = (u_x, u_y, u_z) \quad \text{or} \quad U = \begin{pmatrix} u_x \\ u_y \\ u_z \end{pmatrix}. \tag{II, 1-20b}$$

The magnitude property of such a vector is accounted for by means of the position of the terms in the matrix, the first representing the x direction, the second the y, and the third the z direction. The indices 1, 2, and 3 of (II, 1-19) may be replaced by the coordinates x, y and z of the Cartesian coordinate system and it is seen that, associated with the three eigenvalues (λ) of the matrix, are three distinct characteristic vectors $\mathbf{u}_1$, $\mathbf{u}_2$, $\mathbf{u}_3$. Matrix algebra may also be introduced in multiplying vectors. The scalar product of the vectors **u** and **v** is:

$$\mathbf{u}\cdot\mathbf{v} = u_x v_x + u_y v_y + u_z v_z \tag{II, 1-21a}$$

or in matrix notation:

$$UV^T = (u_x u_y u_z) \begin{pmatrix} v_x \\ v_y \\ v_z \end{pmatrix} = V \cdot U^{-1}. \tag{II, 1-21b}$$

The notation for V^T may be obtained from the definition of symmetrical and antisymmetrical matrices (see Equation (II, 1-18a)). The Equations (II, 1-20) and (II, 1-21) are mentioned here because of their use in the formulation of the algebra of tensors.

Apart from the scalar and cross products of vectors, there is another notation for two vectors **u** and **v**. A *dyad* is represented in vector notation as **uv** and the sum of two or more dyads is called a dyadic. For the dyad **uv** the rules of vector algebra are applicable; thus

$$\begin{aligned} \mathbf{uv} &= (\mathbf{i}u_x + \mathbf{j}u_y + \mathbf{k}u_z)(i\mathbf{v}_x + \mathbf{j}v_y + \mathbf{k}v_z) \\ &= \mathbf{ii}u_x v_x + \mathbf{ij}u_x v_y + \mathbf{ik}u_x v_z + \mathbf{ji}u_y v_x + \mathbf{jj}u_y v_y + \mathbf{jk}u_y v_z + \\ &\quad + \mathbf{kj}u_z v_x + \mathbf{kj}u_z v_y + \mathbf{kk}u_z v_z . \end{aligned} \tag{II, 1-22a}$$

The dyads **ii**, **jj** and **kk** are unit dyads. If **i**, **j** and **k** of (II, 1-22a) are assigned to correspond to positions in an array (first letter refers to the row and the second symbol to the column), then the dyad **uv** becomes:

$$\mathbf{uv} \equiv \begin{pmatrix} \mathbf{ii}u_x v_x & \mathbf{ij}u_x v_y & \mathbf{ik}u_x v_z \\ \mathbf{ji}u_y v_x & \mathbf{jj}u_y v_y & \mathbf{jk}u_y v_z \\ \mathbf{ki}u_z v_x & \mathbf{kj}u_z v_y & \mathbf{kk}u_z v_z \end{pmatrix}. \qquad \text{(II, 1-23a)}$$

A simpler notation is:

$$\mathbf{uv} \equiv \begin{pmatrix} u_x v_x & u_x v_y & u_x v_z \\ u_y v_x & u_y v_y & u_y v_z \\ u_z v_x & u_z v_y & u_z v_z \end{pmatrix} \qquad \text{(II, 1-23b)}$$

where the subscripts x, y and z have taken on the meaning of **i**, **j** and **k**, and the first subscript refers to the row and the second to the column. Equation (II, 1-23b) is the matrix representation of the dyad. It is also the matrix representation of the linear vector function. This function is an operator which produces a vector when it multiplies a vector; an important aspect of this will be treated later on in this chapter. Most important here is the fact that (II, 1-23b) is the *matrix notation* of a second-order tensor in three-dimensional place. The vectors **u** and **v** are represented in matrix notation by

$$\mathbf{u} \equiv U = \begin{pmatrix} u_x \\ u_y \\ u_z \end{pmatrix} \quad \text{and} \quad \mathbf{v} \equiv V = \begin{pmatrix} v_x \\ v_y \\ v_z \end{pmatrix} \qquad \text{(II, 1-24)}$$

and (see Equation (II, 1-21b)), the dyad $\mathbf{D} = \mathbf{uv}$ is thus also represented *in matrix algebra* by the quantity:

$$\mathbf{D} = UV^T = (u_x, u_y, u_z) \begin{pmatrix} v_x \\ v_y \\ v_z \end{pmatrix}. \qquad \text{(II, 1-25)}$$

A dot product may also be taken between a vector **w** and the dyad **uv**. Following through with the rules of vector multiplication, it may be shown that

$$\mathbf{w} \cdot \mathbf{uv} = (\mathbf{w} \cdot \mathbf{u})\, \mathbf{v}$$

and

$$\mathbf{uv} \cdot \mathbf{w} = \mathbf{u}\,(\mathbf{v} \cdot \mathbf{w}) = (\mathbf{v} \cdot \mathbf{w})\, \mathbf{u} \qquad \text{(II, 1-26)}$$

where $\mathbf{u} \cdot \mathbf{v}$ etc., are ordinary scalar products. Tensors of higher order may be defined in the following way. Consider the three vectors **u**, **v** and **w** and by extension of the introduction of the concept of a dyad – let there be defined the quantity

$$\mathbf{uvw}. \qquad \text{(II, 1-27)}$$

There is no rule prohibiting the formulation of a dot product of this quantity and a vector $\mathbf{r}_1$. Thus, (see (II, 1-26))

$$\mathbf{uvw} \cdot \mathbf{r}_1 = (\mathbf{w} \cdot \mathbf{r}_1)\, \mathbf{uv}. \qquad \text{(II, 1-28a)}$$

Now the right member is the product of simply a scalar and the dyad **uv**. Furthermore, by introducing the vector $\mathbf{r}_2$:

$$(\mathbf{w}\cdot\mathbf{r}_1)\,\mathbf{uv}\cdot\mathbf{r}_2 = (\mathbf{w}\cdot\mathbf{r}_1)\,(\mathbf{v}\cdot\mathbf{r}_2)\,\mathbf{u}\,. \qquad \text{(II, 1-28b)}$$

The right member is now a scalar times a vector, and finally:

$$(\mathbf{w}\cdot\mathbf{r}_1)\,(\mathbf{v}\cdot\mathbf{r}_2)\,(\mathbf{u}\cdot\mathbf{r}_3) = \text{scalar}\,. \qquad \text{(II, 1-28c)}$$

II-2. Scattering Tensors and Radiation from Classical Oscillators

In Chapter I the existence of a tensor was indicated which relates the direction of the induced dipole moments and the direction of polarization of the electric field of the incident radiation. There is a similarity in explaining the strain or stress tensor – see the beginning of this chapter – and the scattering tensor. In the latter case, the direction of the induced moment **M** of the scattering entity is not necessarily in the direction of polarization of the incident electric field vector **E**.

The dyad **ME** is represented *in matrix notation* by (see II, 1-23b):

$$\mathbf{ME} \equiv \begin{pmatrix} M_xE_x & M_xE_y & M_xE_z \\ M_yE_x & M_yE_y & M_yE_z \\ M_zE_x & M_zE_y & M_zE_z \end{pmatrix} = \begin{pmatrix} \alpha_{xx} & \alpha_{xy} & \alpha_{xz} \\ \alpha_{yx} & \alpha_{yy} & \alpha_{yz} \\ \alpha_{zx} & \alpha_{zy} & \alpha_{zz} \end{pmatrix} = \boldsymbol{\alpha}\,. \qquad \text{(II, 2-1)}$$

In employing matrix algebra, the tensor α is also defined by (see (II, 1-25)):

$$\boldsymbol{\alpha} = \mathbf{ME}^T = (M_xM_yM_z)\begin{pmatrix} E_x \\ E_y \\ E_z \end{pmatrix}. \qquad \text{(II, 2-2)}$$

Comparing Equations (II, 2-1) and (II, 2-2) it is seen that:

$$\mathbf{M} = \boldsymbol{\alpha}\mathbf{E} \qquad \text{(II, 2-3)}$$

where M, α and E are all given in matrix notation.

Equation (II, 2-3) may be compared to Equation (I, 6-7a) and (I, 6-8a) in Chapter I. Thus the dyad **ME** represents a tensor with the elements $\alpha_{\varrho\sigma}$, and the dot or scalar product of this tensor with the electric field vector **E** generates the induced moment vector **M**. In the specific case where the elements of the tensor α are all real – a symmetric tensor (see Chapter I) – certain characteristic directions are associated with it. The vectors which are associated with these characteristic directions are called proper vectors or eigenvectors, and if these principal vectors are taken as new axes, then the tensor $\boldsymbol{\alpha}$ is written as:

$$\boldsymbol{\alpha} = \begin{pmatrix} \lambda_1 & & 0 \\ & \lambda_2 & \\ 0 & & \lambda_3 \end{pmatrix}. \qquad \text{(II, 2-4)}$$

Here λ_1, λ_2 and λ_3 are certain eigenvalues and they are related to (II, 2-1) and (II, 2-4) in the following way:

$$\begin{pmatrix} \alpha_{xx} & \alpha_{xy} & \alpha_{xz} \\ \alpha_{yx} & \alpha_{yy} & \alpha_{yz} \\ \alpha_{zx} & \alpha_{zy} & \alpha_{zz} \end{pmatrix} \begin{pmatrix} v_x \\ v_y \\ v_z \end{pmatrix} = \begin{pmatrix} \lambda_1 & & 0 \\ & \lambda_2 & \\ 0 & & \lambda_3 \end{pmatrix} \begin{pmatrix} v_x \\ v_y \\ v_z \end{pmatrix} \tag{II, 2-5}$$

where v_x, v_y and v_z are the principal vectors and λ_1, λ_2 and λ_3 are called principal eigenvalues (or principal moments).

Taking the dot product of the vector **E** and the moment **M**

$$\mathbf{E}\cdot\mathbf{M} = \mathbf{E}\cdot\boldsymbol{\alpha}\mathbf{E} \tag{II, 2-6}$$

and assuming that the principal vectors are the axes of the coordinate system, (II, 2-6) can also be written as

$$E_x^2\lambda_1 + E_y^2\lambda_2 + E_z^2\lambda_3 = \mathbf{E}\cdot\mathbf{M} \tag{II, 2-7}$$

because E_x, E_y and E_z are now the components of **E** along these axes. The result is that:

$$\frac{E_x^2}{(\mathbf{E}\cdot\mathbf{M})/\lambda_1} + \frac{E_y^2}{(\mathbf{E}\cdot\mathbf{M})/\lambda_2} + \frac{E_z^2}{(\mathbf{E}\cdot\mathbf{M})/\lambda_3} = 1\,. \tag{II, 2-8}$$

Equation (II, 2-8) is an ellipsoid and by putting **EM** equal unity the expression for the (commonly called polarizability) ellipsoid becomes:

$$\frac{E_x^2}{1/\lambda_1} + \frac{E_y^2}{1/\lambda_2} + \frac{E_z^2}{1/\lambda_3} = 1 \tag{II, 2-9}$$

and the semiaxes are $1/\lambda_1$, $1/\lambda_2$ and $1/\lambda_3$. In the situation where the three principal eigenvalues are equal, the polarizability tensor is spherical symmetric (isotropic) and is automatically diagonal (all $\alpha_{\varrho\sigma}=0$ where ϱ and σ stand for x, y or z) in *any* Cartesian system. If two of the principal moments (λ_1 and λ_2) are equal, then there is cylindrical symmetry about the λ_3-axis. Also, in this situation it is found that any two orthonormal vectors perpendicular to the λ_3-axis are proper principal vectors. In general, however, the scattering tensor is not symmetrical, while in some particular situations the elements might even be complex. A tensor is called *antisymmetric* if $\alpha_{xx}=\alpha_{yy}=\alpha_{zz}=0$ and $\alpha_{xy}=-\alpha_{yx}$; $\alpha_{yz}=-\alpha_{zy}$ and $\alpha_{zx}=-\alpha_{xz}$. The rules of matrix algebra are applicable to tensors and thus, following (II, 1-19); the antisymmetric tensor:

$$\begin{pmatrix} 0 & \alpha_{xy} & \alpha_{xz} \\ -\alpha_{yx} & 0 & \alpha_{yz} \\ -\alpha_{zx} & -\alpha_{zy} & 0 \end{pmatrix} \tag{II, 2-10a}$$

has eigenvalues:

$$\begin{aligned} &\lambda_1 = 0 \\ &\lambda_{2,3} = \pm i\sqrt{\alpha_{xy} + \alpha_{yz} + \alpha_{xz}} \\ &\text{if } |\alpha_{xy}| = |\alpha_{yx}|, \quad |\alpha_{yz}| = |\alpha_{zy}| \text{ and } |\alpha_{xz}| = |\alpha_{zx}|\,. \end{aligned} \tag{II, 2-10b}$$

If ϱ and σ stand for the Cartesian coordinates x, y and z, then an element of the scattering tensor is obtained in the following way:

$$\alpha_{\varrho\sigma} = \alpha^0_{\varrho\sigma}\delta_{\varrho\sigma} + \alpha^s_{\varrho\sigma} + \alpha^a_{\varrho\sigma} \qquad \text{(II, 2-11)}$$

where

$$\alpha^0 = \tfrac{1}{3}(\alpha_{xx} + \alpha_{yy} + \alpha_{zz}) \qquad \text{(II, 2-11a)}$$

and $\delta_{\varrho\sigma}$ is the Kronecker delta and satisfies the relations

$$\delta_{\varrho\sigma} = 0 \quad \text{for} \quad \varrho \neq \sigma$$
$$\delta_{\varrho\sigma} = 1 \quad \text{for} \quad \varrho = \sigma .$$

Furthermore:

$$(\alpha_{\varrho\sigma})^s = \tfrac{1}{2}(\alpha_{\varrho\sigma} + \alpha_{\sigma\varrho}) - \alpha^0\delta_{\varrho\sigma}$$

and

$$(\alpha_{\varrho\sigma})^a = \tfrac{1}{2}(\alpha_{\varrho\sigma} - \alpha_{\sigma\varrho}) . \qquad \text{(II, 2-11b)}$$

The elements $\alpha^s_{\varrho\sigma}$ and $\alpha^a_{\varrho\sigma}$ are elements of the second and third member of (II, 2-10). These tensors are symmetrical and antisymmetrical, respectively. The elements $\alpha^0_{\varrho\sigma}$, given by (II, 2-11a), are elements of the first member of (II, 2-11). These elements are all identical and their values are equal to $\frac{1}{3}$ of the trace of the original tensor. The tensor having the elements $\alpha^0_{\varrho\sigma}$ represents an isotropic tensor. Although anti-symmetric tensors are not encountered very often in physics and chemistry or engineering, they do occur in some Raman effects, and it seems appropriate to pursue these tensors further in more detail. It is important here to take a step back and recall some algebra of dyads. In the preceding paragraph of this chapter it was shown that the dot product of a dyad and a vector results in vector (see Equation (II, 1-26)). Similarly, the question may be asked which dyad **D** satisfies the relation:

$$\mathbf{u} \times \mathbf{v} = \mathbf{D} \cdot \mathbf{v} \qquad \text{(II, 2-12)}$$

In writing out the dot product $D \cdot \mathbf{v}$:

$$\mathbf{D} \cdot \mathbf{v} = \mathbf{D}(\mathbf{i}v_x + \mathbf{j}v_y + \mathbf{k}v_z) . \qquad \text{(II, 2-12a)}$$

On the other hand, the cross product of the vectors **u** and **v** in determinantal form is given by (see II, 1-15)

$$\mathbf{u} \times \mathbf{v} = \begin{vmatrix} \mathbf{i} & \mathbf{j} & \mathbf{k} \\ u_x & u_y & u_z \\ v_x & v_y & v_z \end{vmatrix} . \qquad \text{(II, 2-12b)}$$

By writing out the determinant (see also (II, 1-14)) and comparing components with Equation (II, 2-12a), the dyad **D** is found as:

$$\mathbf{D} = \begin{pmatrix} 0 & u_z & u_y \\ -u_z & 0 & u_x \\ -u_y & -u_x & 0 \end{pmatrix} . \qquad \text{(II, 2-12c)}$$

This tensor or dyad is antisymmetric. In discussing the radiation field associated with an oscillating magnetic dipole – see Chapter I – the conclusion was reached that such a dipole may be formulated by the cross product of two vectors. As a result, it has become customary to associate the scattered radiation from an antisymmetric tensor with the radiation of an oscillating magnetic dipole. Thus for a tensor

$$\begin{pmatrix} 0 & \alpha_{yx} & 0 \\ -\alpha_{yx} & 0 & 0 \\ 0 & 0 & 0 \end{pmatrix}$$

the (magnetic) dipole moment is along the z-axis and the spatial distribution of the electric field is given by (I, 2-12). Vectors which are cross products of ordinary vectors, are called axial vectors or pseudo vectors, as opposed to ordinary or polar vectors. Assuming that the scattering entity is a molecule, atom or ion, it is useful to remark here that an ordinary vector changes sign upon inversion through the origin, but an axial vector remains unaffected. The behaviour of axial vectors, if submitted to certain symmetry operations, is of paramount importance in arriving at selection rules for antisymmetric vectors. In order to find a similar correspondence as in the above between magnetic dipoles and the antisymmetric part of the total scattering tensor, it is necessary to introduce electric quadrupole radiation if the second member of (II, 2-11) is the only tensor which contributes to the intensity of the scattered light. The radiation field of an oscillating quadrupole has been discussed in Chapter I, but another line of thought will be followed here.

Suppose there is a distribution of electrical charges over space in the vicinity of the origin of a coordinate system. Such a physical situation can occur if the scattering system is a molecule, ion or atom. If these particles are exposed to a static electric field **E** then the electrons experience a potential. Of interest here is the potential energy U.

If an electrical charge of *unit* charge is placed at the *origin*, then the potential at the origin may be defined as V_0 and that on the place x_j, y_j, z_j of the jth electron, with charge $-e_j$ is given by V. The potential energy e_jV can be expanded in a Taylor series:

$$e_jV = e_jV_0 + e_j\left[\left(\frac{\partial V}{\partial x}\right)_0 x_j + \left(\frac{\partial V}{\partial y}\right)_0 y_j + \left(\frac{\partial V}{\partial z}\right) z_j\right] +$$
$$+ \tfrac{1}{2}e_j\left[\left(\frac{\partial V}{\partial x^2}\right)_0 x_jx_j + \left(\frac{\partial^2 V}{\partial x\partial y}\right)_0 x_jy_j + \left(\frac{\partial^2 V}{\partial x\partial y}\right)_0 y_jx_j + \cdots\right] +$$
$$+ \text{higher order terms}. \qquad \text{(II, 2-13)}$$

The partial deviations are taken at the origin. Now, the total potential energy U is obtained from contributions of all electrons:

$$U = \sum_j e_jV = V_0 \sum e_j + \nabla V \cdot \sum_j e_j\mathbf{r}_j + \tfrac{1}{2}\sum_j \left[\nabla\left(\nabla V\right) e_j\mathbf{r}_j\right]\mathbf{r}_j +$$
$$+ \tfrac{1}{6}\sum_j \left\{\nabla\left[\nabla\left(\nabla V \cdot e_j\mathbf{r}_j \cdot \mathbf{r}_j\right)\mathbf{r}_j\right]\right\} + \cdots. \qquad \text{(II, 2-14)}$$

Here $\mathbf{V}$ is the ∇-vector and defined as:

$$\mathbf{V} = \frac{\partial}{\partial x}\mathbf{i} + \frac{\partial}{\partial y}\mathbf{j} + \frac{\partial}{\partial z}\mathbf{k} \qquad \text{(II, 2-14a)}$$

and the vector $\mathbf{r}$ may be decomposed in the ordinary way.

The total potential energy U is a scalar, and the potential energy V is also a scalar. The summation $\sum_j e_j$ is the total charge and the second member of (II, 2-14) should also be a scalar. This is in agreement with the fact that the dot product of the ∇-vector and the vector $\mathbf{r}_j$ is a scalar. The vector $\sum_j e_j \cdot \mathbf{r}_j$ represents the dipole moment. The third member of (II, 2-14) may be written – see Equation (II, 1-28c) – as follows:

$$\tfrac{1}{2}\sum_j \left[\mathbf{\nabla}(\mathbf{\nabla}V)\, e_j\mathbf{r}_j\right]\mathbf{r}_j = \tfrac{1}{2}\sum_j \mathbf{\nabla\nabla}V \cdot Q \qquad \text{(II, 2-15)}$$

where $\mathbf{\nabla\nabla}V$ and Q are dyads or tensors of order two. The dyad Q may be written as:

$$Q = \begin{pmatrix} \sum_j x_jx_je_j & \sum_j x_jy_je_j & \sum_j x_jz_je_j \\ \sum_j y_jx_je_j & \sum_j y_jy_je_j & \sum_j y_jz_je_j \\ \sum_j z_jx_je_j & \sum_j z_jy_je_j & \sum_j z_jz_je_j \end{pmatrix} \quad \text{or} \quad Q = \sum_j e_j\mathbf{r}\cdot\mathbf{r}. \qquad \text{(II, 2-16)}$$

Q is the quadrupole tensor and is symmetric. This tensor may also be compared with the symmetric tensor $\alpha^s_{\varrho\sigma}$ (second member) of (II, 2-10), and as a result a similarity is drawn between the radiation field of an oscillating quadrupole – discussed in Chapter I – and that of a symmetric scattering tensor.*

Thus the magnitude of the E-field in space for the quadrupole zz may be computed as follows: if 2δ is the distance of the middle point of the two dipoles (see Figure II-3) and 2ε the distance of the charges of the individual dipoles then

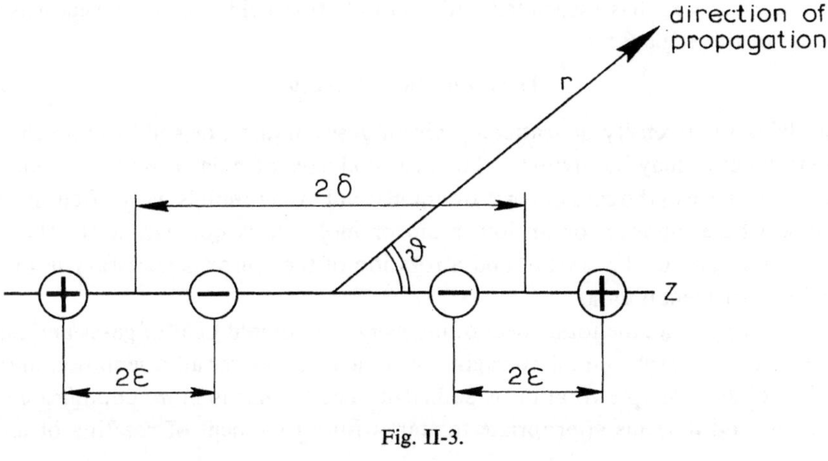

Fig. II-3.

* Note that the experimental conditions should be such that all elements of $\alpha^s_{\varrho\sigma}$ contribute only.

$$Q = \sum_j e_j \mathbf{r}_j \mathbf{r}_j = \sum_j e_j z_j z_j = e[2(\delta + \varepsilon)^2 - 2(\delta - \varepsilon)^2] = 8\, e\varepsilon\delta \qquad \text{(II, 2-17)}$$

and the dipole moment $\mathbf{p} = 2\varepsilon e$, so that

$$Q = 4\delta\mathbf{p}\,. \qquad \text{(II, 2-18)}$$

If the two dipoles oscillate with equal amplitudes but in opposite directions, then the net dipole moment will still be zero but the quadrupole moment varies with time. If p_0 is the distance of the dipoles at the equilibrium position, then it can be shown that:

$$|\mathbf{E}| = Q_0 \frac{4\pi v_0^3}{rc^3} \sin\theta \cos\theta \sin 2\pi v_0 \left(t - \frac{r}{c}\right) \qquad \text{(II, 2-19)}$$

and the spatial distribution of the radiation field vector **E** is that of the familiar d_{z^2} orbital. For the tensor

$$\begin{pmatrix} 0 & \alpha_{xy} & 0 \\ \alpha_{xx} & 0 & 0 \\ 0 & 0 & 0 \end{pmatrix}$$

the radiation field is that of the quadrupole $Q = \sum_j e_j x_j y_j$ and this field is given in Figure I-5. The other three possible quadrupoles have radiation fields which are comparable to the spacial orientation of the $d_{x^2y^2}$, d_{xz} and d_{yz} orbitals. At first glance it seems a bit strange that the functions z^2, $x^2 - y^2$, xy, xz, zy, play such a prominent role here. In Chapter III is shown that these functions have peculiar symmetries and it is the symmetry which dictates the format of the scattering tensors.

Finally the isotropic tensor

$$\begin{pmatrix} \alpha_{xx} & & 0 \\ & \alpha_{yy} & \\ 0 & & \alpha_{zz} \end{pmatrix},$$

where $\alpha_{xx} = \alpha_{yy} = \alpha_{zz}$ has associated with it a radiation field and the **E** vector is independent of the angle θ.*

II-3. Rotation of Tensors

In studying the intensity of scattered light of gases, liquids, or solids, more than one scattering entity may be involved. The principal axes of polarizability are randomly oriented in space in the case of a gas or liquids; but even in solids, very often situations do arise where the axis for an ion, atom or molecule is not parallel to the major macroscopic axes of the crystal, and a rotation of the tensor is necessary in order to explain experimental data.

In order to predict the total effect of intensity of scattered light of gases and liquids, it is necessary to carry out an averaging of some tensors over all orientation in space; but for solids, a simple rotation is sufficient. The former is more complicated than the latter, and it seems appropriate to start with a treatment of rotating of tensors.

* Note that in order that all $\alpha_{\varrho\sigma}$'s $\neq 0$ it is necessary to use incident polarized radiation propagating along more then one of the Cartesian axis!

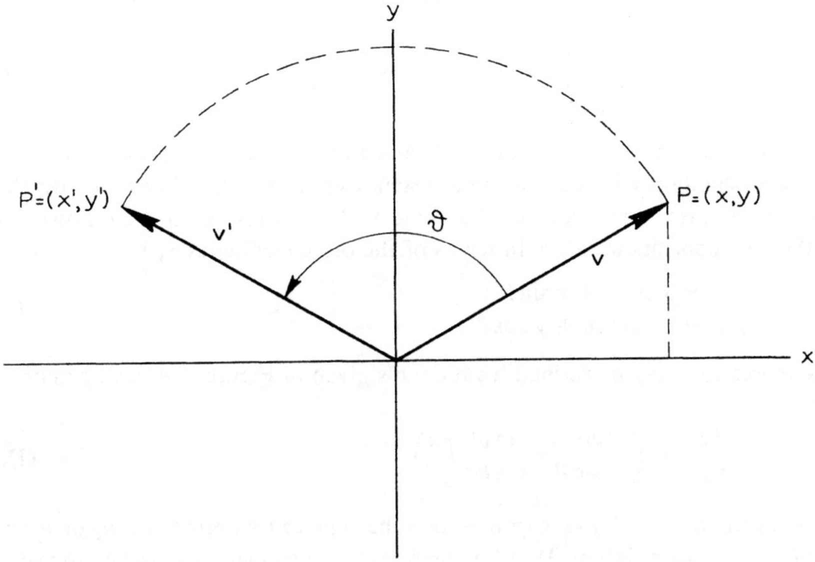
y
P'=(x',y')
P=(x,y)
v'
v
ϑ
x

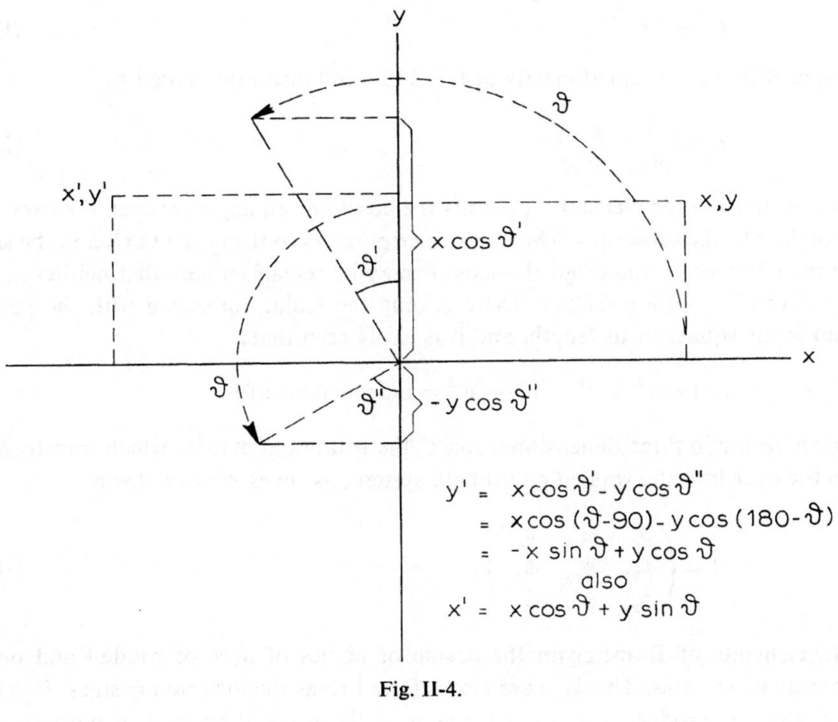
y
ϑ
x',y'
x,y
x cos ϑ'
ϑ'
x
ϑ
ϑ''
- y cos ϑ''
y' = x cos ϑ' - y cos ϑ''
= x cos (ϑ-90) - y cos (180-ϑ)
= - x sin ϑ + y cos ϑ
also
x' = x cos ϑ + y sin ϑ

Fig. II-4.

In order to understand this procedure it is convenient to return to properties of vectors.

The two vectors **u** and **v** may be represented by the following column matrices:

$$\mathbf{u} \equiv U = \begin{pmatrix} u_x \\ u_y \\ u_z \end{pmatrix} \quad \text{and} \quad \mathbf{v} \equiv V = \begin{pmatrix} v_x \\ v_y \\ v_z \end{pmatrix}. \tag{II, 3-1}$$

The vectors are related to a Cartesian coordinate system. Of prime interest here is the transformation which carries this system over to another. Starting with the two-dimensional case, a rotation of θ^0 around the z-axis results in the following values for the new coordinates x', y' in terms of the old coordinates x, y.

$$\begin{aligned} x' &= x \cos\theta + y \sin\theta \\ y' &= -x \sin\theta + y \cos\theta . \end{aligned} \tag{II, 3-2a}$$

This expression may be verified from details given in Figure II-4. Using matrix notation:

$$\begin{pmatrix} x' \\ y' \end{pmatrix} = \begin{pmatrix} \cos\theta & \sin\theta \\ -\sin\theta & \cos\theta \end{pmatrix} \begin{pmatrix} x \\ y \end{pmatrix}. \tag{II, 3-2b}$$

The coordinates x', y' and x, y may now be replaced by either u_x, u_y or v_y, v_x. The mathematical formulation (II, 3-2b) represents the transformation law for vectors in two-dimensional space and can also be written as:

$$U' = \hat{R}U \tag{II, 3-3}$$

where R is the rotational matrix of (II, 3-2b) and also represented by

$$\hat{R} = \begin{pmatrix} \phi_{xx} & \phi_{xy} \\ \phi_{yx} & \phi_{yy} \end{pmatrix}. \tag{II, 3-4}$$

In this matrix each element represents the cosine of an angle between the axes corresponding to the subscripts. The first subscript refers to the primed axis and the second to the unprimed – the original – axis. It may be remarked here that neither x, y nor x', y' of (II, 3-2) are scalars. There is only the scalar connected with the vector **u**, that is the square of its length, and it is easily seen that:

$$|\mathbf{u}^2| = x^2 + y^2 = x'^2 + y'^2 = \text{independent of } \theta .$$

For a vector in three-dimensional space, the rotational matrix, which transforms the vector over into the rotated coordinate system, assumes now the form

$$\hat{R} = \begin{pmatrix} \phi_{xx} & \phi_{xy} & \phi_{xz} \\ \phi_{yx} & \phi_{yy} & \phi_{yz} \\ \phi_{zx} & \phi_{zy} & \phi_{zz} \end{pmatrix}. \tag{II, 3-5}$$

The elements of R are again the cosine of angles of axes of rotated and original coordinate systems. The ϕ_{lk}'s are also referred to as the direction cosines. It is found that the dot product $\mathbf{u}\cdot\mathbf{u}$ is independent of the orientation of **u** in physical space.

The above may now be applied to rotations of scattering tensors because this tensor (or dyad) is given by the quantity α where;

$$\boldsymbol{\alpha} = \mathbf{ME}. \tag{II, 3-6}$$

It is also possible to employ matrix notation. The tensor is then defined by

$$\boldsymbol{\alpha} = \mathbf{ME}^T$$

where $\mathbf{M}$ and $\mathbf{E}^T$ stand for the column and row matrices:

$$\mathbf{M} = (M_x M_y M_z) \quad \text{and} \quad \mathbf{E}^T = \begin{pmatrix} E_x \\ E_y \\ E_z \end{pmatrix}. \tag{II, 3-7}$$

The vectors $\mathbf{M}$ and $\mathbf{E}$ may now be transformed from one coordinate system to another and again, using matrix notation:

$$\mathbf{M}' = \hat{R}\mathbf{M}$$

and

$$\mathbf{E}' = \hat{R}\mathbf{E} \tag{II, 3-8a}$$

where R is the matrix given by (II, 3-4). From the transpose product rule of matrices it follows that:

$$(\mathbf{E}')^T = \mathbf{E}^{-1}\hat{R}^T \tag{II, 3-8b}$$

and

$$\mathbf{M}'(\mathbf{E}')^T = \hat{R}\mathbf{M}\mathbf{E}^T\hat{R}^T \tag{II, 3-8c}$$

or

$$\boldsymbol{\alpha}' = \hat{R}\boldsymbol{\alpha}\hat{R}^T. \tag{II, 3-8d}$$

Here, $\boldsymbol{\alpha}'$ is the tensor in the rotated coordinate system. The elements of the rotational matrix R are all real, and it may directly be shown that

$$\hat{R}^T \cdot \hat{R} = E. \tag{II, 3-9}$$

Following the definition of matrices it is found that the transpose matrix R^T is equal to the inverse matrix $\hat{R}^{-1}$. The matrix $\hat{R}$ is thus orthogonal, and for the elements of the matrices $\hat{R}$ and $\hat{R}^{-1}$ the following relation holds:

$$(\hat{R}^{-1})_{nm} = \hat{R}_{mn}. \tag{II, 3-10}$$

By substituting the elements of the $\hat{R}$ and $\hat{R}^{-1}$ matrix in (II, 3-8d), the tensor α' may be computed; the results for some of the elements of α' are:

$$\begin{aligned} \alpha'_{xx} &= \alpha_{xx}\phi_{xx}^2 + \alpha_{xy}\phi_{xy}\phi_{xx} + \alpha_{xz}\phi_{xz}\phi_{xx} + \alpha_{yx}\phi_{xx}\phi_{xy} + \alpha_{yy}\phi_{xy}^2 + \\ &\quad + \alpha_{yz}\phi_{xz}\phi_{xy} + \alpha_{zx}\phi_{xx}\phi_{xz} + \alpha_{zy}\phi_{xy}\phi_{xz} + \alpha_{zz}\phi_{xz}^2 \\ \alpha'_{xy} &= \alpha_{xx}\phi_{yx}\phi_{xx} + \alpha_{xy}\phi_{yy}\phi_{xx} + \alpha_{xz}\phi_{yz}\phi_{xx} + \alpha_{yx}\phi_{yx}\phi_{xy} + \\ &\quad + \alpha_{yy}\phi_{yy}\phi_{xy} + \alpha_{yz}\phi_{yz}\phi_{xy} + \alpha_{zx}\phi_{yx}\phi_{xz} + \alpha_{zy}\phi_{yy}\phi_{xz} + \alpha_{zz}\phi_{yz}\phi_{xz}. \end{aligned} \tag{II, 3-11}$$

If the tensor $\boldsymbol{\alpha}$ is symmetric, e.g. $\alpha_{xy}=\alpha_{yx}$, $\alpha_{yz}=\alpha_{zy}$ and $\alpha_{xz}=\alpha_{zx}$, then the following general expression may be derived for an element α'_{np} of the transformed tensor α':

$$\begin{aligned}\alpha'_{np} &= \sum_{kl} \hat{R}_{nk}\alpha_{kl}(\hat{R}^{-1})_{lp} \\ &= \sum_{kl} \hat{R}_{nk}\hat{R}_{pl}\alpha_{kl}\,.\end{aligned} \qquad \text{(II, 3-12)}$$

If the original tensor $\boldsymbol{\alpha}$ is antisymmetric, e.g. $\alpha_{xy}=-\alpha_{yx}$, $\alpha_{yz}=-\alpha_{zy}$ and $\alpha_{xz}=-\alpha_{zx}$ with $\alpha_{xx}=\alpha_{yy}=\alpha_{zz}=0$, then (II, 3-12) may also be used, but the condition $l\neq k$ should be attached. It may be of interest to discuss now the invariances of a 3×3 tensor. An invariance is a quantity whose value does not change when the coordinate axes are rotated. For the present matrix, these quantities are:

(i) The trace of the tensor. Thus:

$$\alpha_{xx} + \alpha_{yy} + \alpha_{zz} = \alpha'_{xx} + \alpha'_{yy} + \alpha'_{zz} \qquad \text{(II, 3-13a)}$$

(ii) The sum of the principal two-rowed minors:

$$\begin{vmatrix}\alpha_{yy} & \alpha_{yz}\\ \alpha_{zy} & \alpha_{zz}\end{vmatrix} + \begin{vmatrix}\alpha_{xx} & \alpha_{xz}\\ \alpha_{zx} & \alpha_{zz}\end{vmatrix} + \begin{vmatrix}\alpha_{xx} & \alpha_{xy}\\ \alpha_{yx} & \alpha_{yy}\end{vmatrix} = \begin{vmatrix}\alpha'_{yy} & \alpha'_{yz}\\ \alpha'_{zy} & \alpha'_{zz}\end{vmatrix} + \begin{vmatrix}\alpha'_{xx} & \alpha'_{xz}\\ \alpha'_{zx} & \alpha'_{zz}\end{vmatrix} + \begin{vmatrix}\alpha'_{xx} & \alpha'_{xy}\\ \alpha'_{yx} & \alpha'_{yy}\end{vmatrix}$$

(iii) The determinant of the matrix:

$$|\alpha| = |\alpha'|$$

The invariance of, for instance, the trace of the tensors may be proven in the following way, using (II, 3-12):

$$\sum_n \alpha'_{nn} = \sum_n \sum_{kl} \hat{R}_{nk}\hat{R}_{nl}\alpha_{kl} = \sum_{kl} \alpha_{kl} \sum_n \hat{R}_{nk}\hat{R}_{nl} = \sum_{kl} \alpha_{kl}\delta_{kl} = \sum_k \alpha_{kk}\,. \qquad \text{(II, 3-14)}$$

This paragraph is concluded with some remarks on the similarity of transformation of elements of the tensor α and that of products of the type x^2, xy etc. where x, y and z are coordinates.

The transformation of the coordinates x, y and z to the primed coordinates x', y' and z' is achieved using the rotation matrix R. Thus:

$$\begin{aligned}x' &= \phi_{xx}x + \phi_{xy}y + \phi_{xz}z\\ y' &= \phi_{yx}x + \phi_{yy}y + \phi_{yz}z\\ z' &= \phi_{zx}x + \phi_{zy}y + \phi_{zz}z\end{aligned} \qquad \text{(II, 3-15)}$$

and for the products $(x')^2$, $(x'y')$ etc. the relations are:

$$\begin{aligned}x'^2 = \phi_{xx}^2x^2 &+ \phi_{xy}\phi_{xx}xy + \phi_{xz}\phi_{xx}xz + \phi_{xy}\phi_{xx}yx + \phi_{xy}^2y^2 + \\ &+ \phi_{xz}\phi_{xy}yz + \phi_{xx}\phi_{xz}zx + \phi_{xy}\phi_{xz}zy + \phi_{xz}^2z^2\,.\end{aligned} \qquad \text{(II, 3-16)}$$

It may thus be concluded from (II, 3-16) and (II, 3-11) that α_{xx} transforms like x^2, α_{xy} like the product xy. These similarities are quite useful if the selection rules for transitions are derived, in the same way as in spectroscopy the existence of equivalences of p_x, p_y and p_z-orbitals and the coordinates x, y and z are utilized. For the antisymmetric tensors like

$$\begin{pmatrix} 0 & \alpha_{xy} & 0 \\ -\alpha_{yx} & 0 & 0 \\ 0 & 0 & 0 \end{pmatrix}$$

the similarity does not seem to hold because the quantity xy-yx is in general equal to zero. In order to retain the similarity, the fiction has to be adopted that xy and yx are noncommutative.

II-4. Specific Rotation of Symmetric and Antisymmetric Tensors

For a rotation of a tensor α of θ^0 around the z-axis of the Cartesian coordinate system, the $\hat{R}$ matrix is equal to

$$\hat{R} = \begin{pmatrix} \cos\theta & \sin\theta & 0 \\ -\sin\theta & \cos\theta & 0 \\ 0 & 0 & 1 \end{pmatrix} \qquad \text{(II, 4-1)}$$

and the elements of $\hat{R}^{-1}$ are:

$$(R^{-1})_{ij} = R_{ji}. \qquad \text{(II, 4-2)}$$

Hence, it is now possible to construct the rotated tensors using (II, 3-12), assuming that the original tensors are either symmetric or antisymmetric. Some of the examples chosen here given below are important in that they are the tensors which appear in Raman transitions of systems having cubic symmetry. A most convenient scattering tensor to rotate is an isotropic tensor α given by

$$\boldsymbol{\alpha} = \begin{pmatrix} \alpha_{xx} & 0 & 0 \\ 0 & \alpha_{yy} & 0 \\ 0 & 0 & \alpha_{zz} \end{pmatrix} \qquad \text{(II, 4-3)}$$

where $\alpha_{xx}=\alpha_{yy}=\alpha_{zz}$. According to the rule for invariance of tensors (see (II, 3-13a)), it follows that for the diagonal elements α'_{ii} of the rotated tensor α' the relation $\alpha'_{xx}+\alpha'_{yy}+\alpha'_{zz}=\alpha_{xx}+\alpha_{yy}+\alpha_{zz}$ holds. Furthermore, from (II, 3-12) it is possible to find the values of the elements α'_{ij}. The result is that

$$\alpha'_{xx} = \alpha'_{yy} = \alpha'_{zz} = \alpha_{xx} = \alpha_{yy} = \alpha_{zz}$$

and furthermore, all

$$\alpha_{ij} = \alpha'_{ji} = 0.$$

An isotropic tensor remains unaltered if rotated around the z-axis (and also around the y- and x-axes). This is an important result in explaining the polarization properties of scattered light.

Another tensor of interest is

$$\boldsymbol{\alpha} = \begin{pmatrix} a & 0 & 0 \\ 0 & a & 0 \\ 0 & 0 & -2a \end{pmatrix}. \qquad \text{(II, 4-4)}$$

The trace of α equals zero, and a rotation around the z-axis leaves the tensor unaltered. This is not the case, however, if the tensor α is rotated around either the x- or y-axis. This may be demonstrated by using as an example the tensor

$$\boldsymbol{\alpha} = \begin{pmatrix} a & 0 & 0 \\ 0 & -a & 0 \\ 0 & 0 & 0 \end{pmatrix}. \tag{II, 4-5a}$$

A rotation around the z-axis, when the angle of rotation equals $\theta = 45$ results in α' where:

$$\boldsymbol{\alpha}' = \begin{pmatrix} 0 & -a & 0 \\ -a & 0 & 0 \\ 0 & 0 & 0 \end{pmatrix}. \tag{II, 4-5b}$$

The conclusion is reached that some tensors having diagonal elements may be transformed in tensors having diagonal elements with values equal to zero. In certain specific cases it may be possible to follow the value of the element $\alpha'_{\varrho\sigma}$ as a function of the angle of rotation θ. A result is given in Figure II-5, and Equation (II, 4-5a)

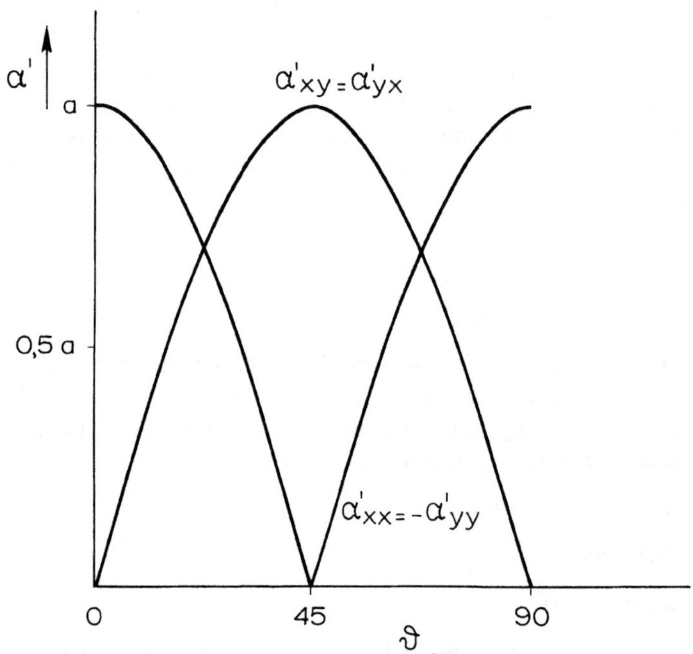

Fig. II-5. Value of the elements α'_{xx} and α'_{xy} of the tensors II, 4-5a, b as function of θ.

describes only one situation for which $\theta = 45°$. A tensor having only diagonal elements is:

$$\boldsymbol{\alpha}^S = \begin{pmatrix} 0 & a & 0 \\ a & 0 & 0 \\ 0 & 0 & 0 \end{pmatrix}. \tag{II, 4-6a}$$

If rotated around the z-axis, and assuming $\theta = 45°$, then:

$$\boldsymbol{\alpha}'^S = \begin{pmatrix} a & 0 & 0 \\ 0 & a & 0 \\ 0 & 0 & 0 \end{pmatrix}. \qquad \text{(II, 4-6b)}$$

The tensors discussed here are all symmetric tensors. A typical example of an antisymmetric tensor is:

$$\boldsymbol{\alpha}^A = \begin{pmatrix} 0 & a & 0 \\ -a & 0 & 0 \\ 0 & 0 & 0 \end{pmatrix}. \qquad \text{(II, 4-7a)}$$

This tensor rotated around the z-axis results in the transformed tensor α'^A. When the angle of rotation equals 45°:

$$\boldsymbol{\alpha}'^A = \begin{pmatrix} 0 & a & 0 \\ -a & 0 & 0 \\ 0 & 0 & 0 \end{pmatrix}. \qquad \text{(II, 4-7b)}$$

It may be possible again to follow the value of the element α'_{xy} of the tensors α'^A and α'^S as a function of the angle of rotation θ. The result is presented in Figure II-6. The figure also indicated that for tensors which are partly antisymmetric, the amount of antisymmetry may be determined from the result of an experiment described by Figure II-6. If the value of an element of α is given by:

$$\alpha = p\alpha^A + q\alpha^S \qquad \text{(II, 4-8)}$$

then the ratio p/q may be evaluated from the θ dependent value of an element of the rotated tensors. The value of this ratio is indicated in Figure II-6. The calculations for values of elements of tensors are more complicated if the scattering entities are distributed in a random orientation of N particles in physical space. In Chapter I it is shown that the intensity of the scattered radiation is always proportional to the square of the induced moments and thus also to the square of the elements of the scattering tensor. Therefore, in order to relate the calculations of averaged values to physical significant properties, it is preferable that the averaged values of the square of the elements of a tensor be considered.

According to (II, 3-12), the value of an element α_{nm} of a transformed symmetrical tensor α is given by:

$$\alpha'_{nm} = \sum_{kl} \hat{R}_{nk}\hat{R}_{ml}\alpha_{kl}. \qquad \text{(II, 4-9a)}$$

It is now convenient to introduce the principal axis of the tensor α. The elements of this tensor – the principal valuec – are called $(\alpha_1, \alpha_2, \alpha_3)$ or α_i, where $i = 1, 2$ or 3 and the position of the principal axis denotek by λ_1, λ_2 and λ_3 respectively is shown in Figure II-7 relative to the laboratory fixed axis X, Y and Z. Expression (II, 4-9a) can then be written as:

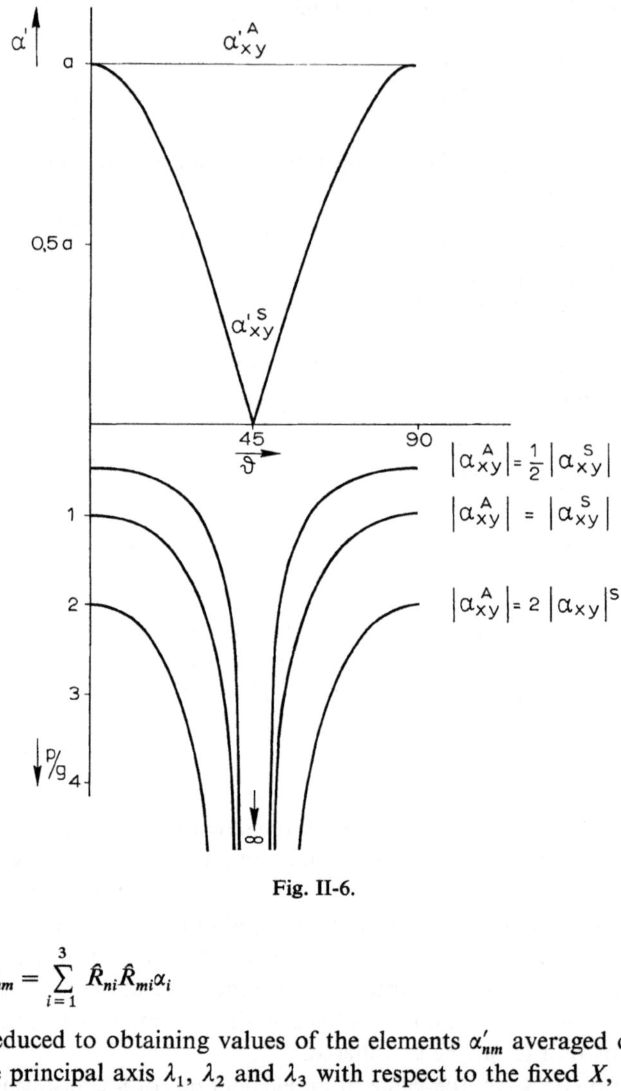

Fig. II-6.

$$\alpha'_{nm} = \sum_{i=1}^{3} \hat{R}_{ni}\hat{R}_{mi}\alpha_i \tag{II, 4-9b}$$

and is now reduced to obtaining values of the elements α'_{nm} averaged over all orientations of the principal axis λ_1, λ_2 and λ_3 with respect to the fixed X, Y and Z axis. The quantities α_i are, of course, constants under the averaging process. For the averaged value of the square of an element α_{nm} the following relation may be written:

$$\begin{aligned}\overline{(\alpha'_{nm})^2} &= \overline{\left(\sum_{i=1}^{3} R_{ni}R_{mi}\alpha_i\right)^2} \\ &= \sum_{i=1}^{3} \alpha_i^2\overline{R_{ni}^2R_{mi}^2} + 2\sum_{i<i'} \alpha_i\alpha_{i'}\overline{R_{ni}R_{mi}R_{ni'}R_{mi'}}\,. \end{aligned} \tag{II, 4-9c}$$

The easiest terms to evaluate are the terms with the indexes $n=m$. The values of R_{in} refer to the cosine of the angle between the n-axis (e.g. X, Y or Z) and the principal axis λ_1, λ_2 and λ_3; the angle between λ_3 and Z is denoted by θ, and the angle of the projection of λ_3 in the XY plane and the Y axis is called φ. The averaged value of α'_{nn} can then be written:

$$\overline{(\alpha'_{nn})^2} = \sum_{i=1}^{3} \alpha_i^2 \overline{R_{ni}^4} + 2 \sum_{i<i'} \alpha_i \alpha_{i'} \overline{R_{ni}^2 R_{ni'}^2} \qquad \text{(II, 4-10a)}$$

where

$$\overline{R_{ni}^4} = \overline{\cos^4 \theta} = \frac{1}{4\pi} \int_0^{\pi} \int_0^{2\pi} \cos^4 \theta \sin \theta \, d\theta \, d\varphi = \tfrac{1}{5}. \qquad \text{(II, 4-10b)}$$

Furthermore, it is directly seen from Figure II-7 that

$$\cos^2 (X, \lambda_3) + \cos^2 (Y, \lambda_3) + \cos^2 (Z, \lambda_3) = 1. \qquad \text{(II, 4-11)}$$

Similar expressions are valid for the cosine of X, Y and Z and λ_1 and λ_2. By squaring and averaging it is found that:

$$\overline{3R_{ni}^4} + \overline{6R_{ni}^2 R_{mi'}^2} = 1. \qquad \text{(II, 4-12a)}$$

Using the results of (II, 4-10b), it follows that:

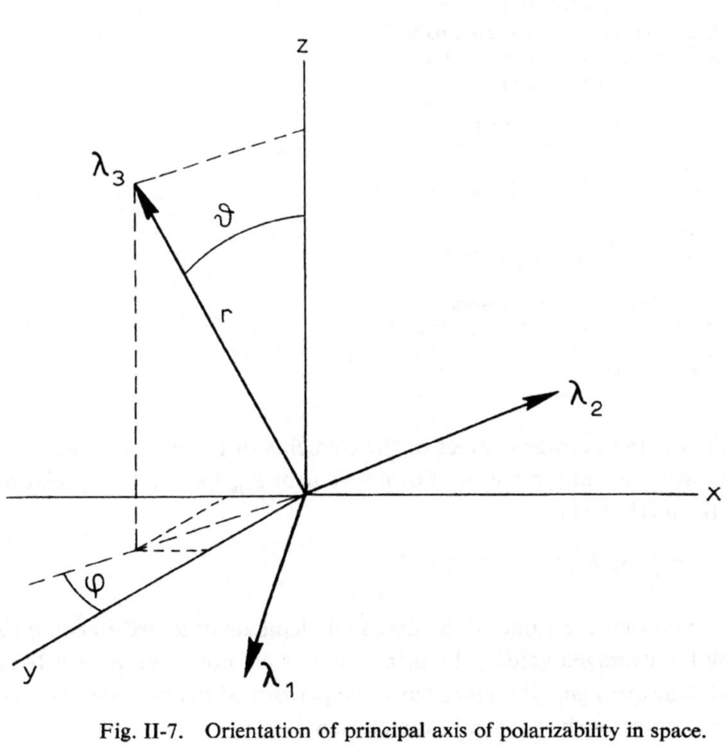

Fig. II-7. Orientation of principal axis of polarizability in space.

$$\overline{R_{ni}^2 R_{mi}^2} = \tfrac{1}{15} \quad \text{but also} \quad \overline{R_{ni}^2 R_{ni'}^2} = \tfrac{1}{15}. \tag{II, 4-12b}$$

Thus

$$\overline{(\alpha'_{nn})^2} = \tfrac{1}{5} \sum_i \alpha_i^2 + \tfrac{2}{15} \sum_{i<i} \alpha^i \alpha^{i'} \tag{II, 4-12c}$$

The results obtained in (II, 4-12b) may be compared directly with the value of the second member of (II, 4-9c) under the condition that $i = i'$ (in (II, 4-9c)).

Returning to Figure II-6 it may be shown that:

$$\cos(X, \lambda_3)\cos(X, \lambda_2) + \cos(Y, \lambda_3)(\cos Y, \lambda_2) + \\ + \cos(Z, \lambda_3)(Z, \lambda_2) = 0 \tag{II, 4-13a}$$

and similarly for other combinations of angles between the X, Y and Z and λ_1, λ_2 and λ_3 axes. Squaring and averaging then results in the following expression:

$$3\overline{R_{ni}^2 R_{ni'}^2} + 6\overline{R_{ni} R_{mi} R_{ni'} R_{mi'}} = 0. \tag{II, 4-13b}$$

In employing Equation (II, 4-12b) the final result is that:

$$\text{for} \quad i \neq i' \quad \overline{R_{ni} R_{mi} R_{ni'} R_{mi'}} = -\tfrac{1}{30}. \tag{II, 4-13c}$$

The results of the computation of the average values of the direction cosine is also tabulated in Table II-1.

TABLE II–1

Elements of the scattering matrix for a system of random oriented scattering centers.

Symmetric Tensors
$\overline{\alpha_{nn}^2} = \frac{1}{5} \sum_i \alpha_i^2 + \frac{2}{15} \sum_{i<i'} \alpha_i \alpha_{i'}$
$\overline{\alpha_{nm}^2} = \frac{1}{15} \sum_i \alpha_i^2 - \frac{1}{15} \sum_{i<i'} \alpha_i \alpha_{i'}$
Antisymmetric Tensors
$\overline{\alpha_{nm}^2} = \frac{1}{10} \alpha_{kl}^2$

In a similar way the averaged values of the elements of the tensor α' may be calculated if the tensor α is antisymmetric. For an element α'_{nm} the following relation may be obtained from (II, 3-11):

$$\alpha'_{nm} = \sum_{kl} R_{nk} R_{ml} \alpha_{kl} \quad \text{for} \quad k \neq l. \tag{II, 4-14a}$$

It should be noted that the value of the diagonal elements of α' are all equal to zero. In calculating the averaged values, the principal axis cannot be employed because it does not exist. The averaging therefore has to be performed maintaining the terms α_{kl}. Thus, for the average of the square of the elements α'_{nm}:

$$\overline{(\alpha'_{nm})^2} = \overline{(\sum_{kl} \phi_{ml}\phi_{nk}\alpha_{kl})^2} \quad \text{for} \quad k \neq l. \tag{II, 4-14b}$$

For an antisymmetric tensor, $\alpha_{kl} = -\alpha_{lk}$ and Equation (II, 4-14a) can also be written as:

$$\overline{(\alpha'_{nm})^2} = \overline{\sum_{kl} (\phi_{ml}\phi_{nk} - \phi_{km}\phi_{ln})^2 (\alpha_{lk})^2}. \tag{II, 4-14c}$$

The conditions for (II, 4-14b) are that $l \neq k$, but there are no restrictions with respect to the indices n and m. The averaged value of α_{lk} is identical to itself and, employing (II, 4-12b) and (II, 4-13b), it may be shown that:

$$\overline{\sum_{kl} (\phi_{ml}\phi_{nk} - \phi_{km}\phi_{ln})^2} = \tfrac{1}{10}. \tag{II, 4-15}$$

Table (II-1) is important in that the averaged values may be used to calculate depolarization ratios for systems of random oriented particles.

The index n and m refer again to the row and column of a matrix. But the intensity of the scattered light is always proportional to the square of an element of the tensor Thus the observed intensities are also related to elements of a matrix, the *scattering matrix*. Thus if the incident light is plane-polarized with its direction of propagation along the Y-axis of Figure II-7, then for the scattered light in the X-direction the intensity I is:

$$I \approx \overline{\alpha_{nn}^2} + \overline{\alpha_{nm}^2} \quad \text{or} \quad \begin{cases} I_z \approx \overline{\alpha_{zz}^2} + \overline{\alpha_{yz}^2} & \text{for } \mathbf{E} = \mathbf{E}_z \\ I_x \approx \alpha_{zx}^2 + \alpha_{yx}^2 & \text{for } \mathbf{E} = \mathbf{E}_x. \end{cases} \tag{II, 4-16}$$

If the vector of the incident light is parallel to the Z-axis and the scattered light is analyzed with a sheet of polaroid so that the direction of preference of the electric vector is also parallel to the Z-axis, then I^z is:

$$I_z^z \approx \overline{\alpha_{zz}^2} \tag{II, 4-16a}$$

and if the direction of preference is parallel the Y-axis, then I^y is:

$$I_z^y \approx \overline{\alpha_{yz}^2}. \tag{II, 4-16b}$$

The above expressions are used in Chapter IV-8 to determine the values of the depolarization factor.

Suggestions for Further Reading

[1] Wilson, E. B., Decius, J. C., and Cross, P. C.: 1955, *Molecular Vibrations*, McGraw-Hill, New York.

CHAPTER III

SOME ASPECTS OF GROUP THEORY

In the discussion offered in Chapters I and II, no mention has been made of the impact of the symmetry of the scattering entity, e.g. atom, ion or molecule, on the properties of the scattered light. Yet, intuitively it can be felt that the properties of, for instance, the Raman radiation of a molecule like CCl_4 could be different from that of the $CHCl_3$, not because a hydrogen atom replaces a chlorine atom of CCl_4 but because the symmetry of the latter molecule is different from that of the former. The symmetry difference is reflected, for example, through the absence of a three-fold rotational axis around any of the C-Cl bonds of the $CHCl_3$ molecule, while four of such rotational axes are present for the CCl_4 molecule. The effect of symmetry on the Rayleigh and Raman processes should, of course, be reflected in the scattering tensor – not to such extent that a separation of symmetric and antisymmetric tensors may occur, (they are associated with the type of Rayleigh and Raman process) – but more importantly, it determines which of the six elements of the symmetric tensor and three elements of the anti-symmetric tensor are different from zero. It may be remarked here that the trend is that the scattering tensor will become more simple (more elements will be equal to zero) if the symmetry of the scattering entity becomes higher. The interesting aspect of application of group theory is that a considerable amount of information on the tensor is obtained without going into details, about the mathematical formulation of the elements of the scattering tensor. In particular, the selection rules for the Raman scattered light are applicable to the above, while for the actual calculation of absolute intensities, emphasis must be placed on the appropriate expression for the said elements. For the purpose of the explanation of the various Raman effects, it is not necessary to have at hand the details of group theory. In this chapter some aspects will be commented upon.

III-1. Symmetry Elements

In the introduction to this chapter, use was made of the special arrangements of atoms of the molecules CCl_4 and $CHCl_3$ to demonstrate difference of symmetry – in particular, the absence of a three-fold rotational axis in the latter arrangement. Such an axis is called a *symmetry element*. These elements are not restricted to the presence of a rotational axis. There are also mirror planes, centra of symmetry (inversion centers), and improper rotational axes. The *symmetry operations* associated with these elements are indicated in Table III-1.

The symmetry elements mentioned here are appropriate in describing the symmetry

TABLE III-1
Symmetry elements and operations

Element	Symbol	Operation	Symbol
Identity	E	Leave atoms unaffected	$\hat{E}$
Rotation axis	$C_n{}^m$	m successive rotations of $2\pi m/n$ degrees	$\hat{C}_n{}^m$
Inversion center	i	Inversion of all atoms through the center	$\hat{i}$
Improper rotation axis	$S_n{}^m$	m successive rotations of $2\pi m/n$ degrees followed by a reflection in a plane $\perp$ to the rotation axis	$\hat{S}_n{}^m$

of free molecules. For atoms or ions in crystals, some additional elements have to be considered. For the latter, e.g. crystals, such symmetry operations as (i) translations and (ii) special rotations which can occur only when combined with special translations (screw and glide operations) should be added to the operations mentioned above. For free atoms or ions (assuming spherical symmetry of the cloud of electrons), there is full rotational symmetry; that is, n-fold rotational axes pointing to the nucleus are distributed over all directions in three-dimensional space. The symmetry operations are denoted by special symbols indicated in the fourth column of Table III-1. The effect of applying the operation $\hat{\sigma}$ produces a configuration which is *equivalent* to the original, but the second application of the same $\hat{\sigma}$ produces a configuration *identical* with the original. These operations may also be represented in the following way:

$$\sigma \cdot \sigma = \sigma^2 = E. \tag{III, 1-1}$$

If the successive operation would have been taken n times, then this may be denoted by σ^n. The symbol E is denoted by the identity operation and it is obvious that when n is even, $\sigma^n = E$; while n is odd, $\sigma^n = \sigma$.

A reflection through a plane of a point in space designated with the coordinates x, y and z, results in the transformed coordinates x', y' and z'. The primed coordinates are, in general, a linear combination of the coordinates x, y and z; but if the plane of reflection contains the original x-and y-axis, then the operation $\hat{\sigma}_{xy}$ leads to the following

$$\begin{pmatrix} x' \\ y' \\ z' \end{pmatrix} = \hat{\sigma}_{xy} \begin{pmatrix} x \\ y \\ z \end{pmatrix} \tag{III, 1-2a}$$

where

$$\hat{\sigma}_{xy} = \begin{pmatrix} 1 & & \\ & 1 & \\ & & -1 \end{pmatrix}. \tag{III, 1-2b}$$

A successive operation as indicated in (III, 1-1) may also be represented in terms of multiplication of matrices; the matrix associated with the identity operation $\hat{E}$ is then

$$\hat{E} = \begin{pmatrix} 1 & & \\ & 1 & \\ & & 1 \end{pmatrix}. \tag{III, 1-3}$$

The effect of carrying out the operation associated with an inversion center, located at the origin of the coordinate system, is that $x' = -x$, $y' = -y$ and $z' = -z$. The matrix associated with such an operation may be written as:

$$\hat{\imath} = \begin{pmatrix} -1 & 0 & 0 \\ 0 & -1 & 0 \\ 0 & 0 & -1 \end{pmatrix}. \qquad \text{(III, 1-3a)}$$

A successive operation brings the system back into its original configuration and, as in the case of the $\hat{\sigma}$ operation, it is found that:

$$\hat{\imath}^n = E \quad \text{for} \quad n = \text{even}$$
$$\hat{\imath}^n = \hat{\imath} \quad \text{for} \quad n = \text{odd}.$$

The transformation matrix of a general point in space under rotation around the Cartesian axis has been discussed in detail in the preceding chapter. In general, an n-fold axis is denoted C_n and a rotation by $2\pi/n$ is also represented by this symbol. Rotations by $2\pi/n$ carried out successively m times are represented by the symbol C_n^m, and for any rotation axis it is found:

$$C_n^n = E \quad \text{and} \quad c_n^{n+1} = C_n, \qquad C_n^{n+2} = C_n^2 \quad \text{etc.} \qquad \text{(III, 1-4)}$$

It is seen that a proper axis of rotation generates n operations, this being contrary to the operations $\hat{\sigma}$ and $\hat{\imath}$ where only one operation for each symmetry element is involved. If, for instance, a C_3-axis and a C_2-axis are perpendicular to each other, then as a direct consequence of the various operations it may be shown that two other C_2-axes must exist perpendicular to C_3, making angles of $2\pi/3$ and $4\pi/3$ with the first two-fold axis. Such a situation does not occur if the C_3-axis is perpendicular to a plane of reflection. Apart from the proper rotation axis discussed above, there are also improper axes. The operation associated with this symmetry element may be broken down into two steps. First a proper rotation is carried out around the axis followed by a reflection through a plane perpendicular to this axis. For example, for the operation $\hat{S}_4^1$, a general point in space with the coordinates (x, y, z) is carried over in x', y', z' where $x' = -y$, $y' = x$ and $z' = -z$. The matrix associated with the operation $\hat{S}_4^1$ is thus:

$$\hat{S}_4^1 = \begin{pmatrix} 0 & -1 & 0 \\ 1 & 0 & 0 \\ 0 & 0 & -1 \end{pmatrix}. \qquad \text{(III, 1-5a)}$$

Using this routine, it is relatively simple to show that

$$\hat{S}_2^2 = \hat{\imath}^2 = E \qquad \text{(III, 1-5b)}$$

and

$$\hat{S}_n^n = \bar{C}_n^n = E \quad (n \text{ is even}) \qquad \text{(III, 1-5c)}$$

or

$$\hat{S}_n^n = \sigma$$

but

$$S_n^{2n} = E. \quad (n \text{ is odd}). \qquad \text{(III, 1-5d)}$$

The improper axis of rotation S_n^n also generates n operations, but some of these operations may be written in a simpler way, as may be demonstrated with the following example. The six operations of the symmetry element S_6 are represented by the operations:

$$\hat{S}_6, \quad \hat{S}_6^2 = \hat{C}_3, \quad \hat{S}_6^3 = \hat{\imath}, \quad \hat{S}_6^4 = \hat{C}_3^2, \quad \hat{S}_6^5, \quad \hat{S}_6^6 = E. \tag{III, 1-5e}$$

The symmetry elements described here are of particular interest in the application of group theory to free atoms, ions and molecules, but in the solid state, there are additional unique symmetry elements.

The application of group theory to crystals mainly centers around the symmetry properties of the unit cell of these crystals. The axis of these cells may be denoted by the basic vectors $\mathbf{a}_x$, $\mathbf{a}_y$ and $\mathbf{a}_z$ so that the position vector $\mathbf{r}$ of any lattice point can be expressed by:

$$\mathbf{r} = n_x\mathbf{a}_x + n_y\mathbf{a}_y + n_z\mathbf{a}_z\,. \tag{III, 1-6}$$

The unit cell of high symmetry lattices is such that the basic vectors are perpendicular (orthogonal), but for low symmetry lattices this is not necessarily so. The coefficients n_x, n_y and n_z are all integers. As a result of a primitive lattice operation $\hat{t}$, a vector

$$\hat{t} = t_x\mathbf{a}_x + t_y\mathbf{a}_y + t_z\mathbf{a}_z \tag{III, 1-7}$$

is added to every position vector $\mathbf{r}$ given by (III, 1-6). The difference of this operation $\hat{t}$ and any of $\hat{E}$, $\hat{\sigma}$, $\hat{\imath}$, $\hat{C}_n^m$ and $\hat{S}_n^m$ is that $\hat{t}$ does not leave the origin of the coordinate system invariant, and it is not possible to have the operation represented by a matrix.

There are two nonprimitive translations, the glide plane and screw axis respectively. The operation associated with the former symmetry element may be broken down into two steps. A point in the unit cell is reflected through a plane followed by a nonprimitive translation over a distance $(1/n_x)a_x$. Such an operation may be represented by a matrix, but the elements of this matrix can only be computed if the location of the plane and the distance $(1/n_i)a_i$ inside the unit cell are known. This is also valid for the screw axis. The operations corresponding to this symmetry element are rotations around a proper axis followed by a nonprimitive translation parallel to this axis.

III-2. Definition Properties of Point, Space and Factor Groups

A group is defined as a collection of elements. Special laws exist for combining two elements, and the laws may be formulated as follows:

(I) A combination of any two elements (A and B) of the group must result in another element of this group. This may be represented as

$$AB = C\,. \tag{III, 2-1a}$$

(II) The associate law of combining elements must hold. Thus, if A, B and C are elements of the group, then:

$$A(BC) = (AB)\,C\,. \qquad \text{(III, 2-1b)}$$

(III) There always exists in a group a unique element called the identity element E. The following relation holds:

$$EA = AE = A\,. \qquad \text{(III, 2-1c)}$$

(IV) Associated with every element (A) of a group is one, and only one, inverse element A^{-1} so that

$$A^{-1}A = AA^{-1} = E\,. \qquad \text{(III, 2-1d)}$$

It follows from the above that a group can have a finite or an infinite number of elements, and the order of the group is equal to the number of elements. If the law of combining of elements obeys the relation $AB = BA = C$, then it is said that the elements A and B commute, and the group is called an Abelian group.

A rather broad meaning should be attached to the operation 'combining of elements' For instance, a product is one type of combination; addition, subtraction and also differentiation are other examples. If the elements are simply all integer numbers (positive, negative and zero), then groups do exist where the types of combination are multiplication, addition or subtraction. These groups are Abelian and they are also examples of infinite groups. Both infinite and finite groups play an important role in explaining the properties of scattered light. The elements which form the basis of these groups are the symmetry elements discussed in Section III-1. Exclusive of the primitive and nonprimitive translations, 32 point groups may be constructed using combinations of the symmetry elements C_n^m, σ, i, S_n^m and E. All these point groups are Abelian and two are infinite. If the primitive translations are added, another 73 groups are obtained. Finally, after introducing the nonprimitive translations, this gives 157 groups. There are a total of $73 + 157 = 230$ space groups.

The selection of symmetry elements as a first step to construct a group is not done at random. There are rules which have to be taken into account and they may be explained using the following example. Suppose the symmetry element E, C_4 and σ are chosen. The location of the four-fold axis and the plane are indicated in Figure III-1. There are the following operations associated with the three symmetry elements: $\hat{E}$, $\hat{C}_4^3$, $\hat{C}_4^2 = \hat{C}_2$ and $\hat{\sigma}$. But, as a result of these operations, three more planes are generated if the σ plane contains the z- and x-axis. One of these planes is perpendicular to σ_{zx} and contains the z- and y-axis, but there are also two planes which are indicated in the figure and denoted σ_d^1 and σ_d^2. Thus, the total number of symmetry elements is eight, but not all of them are relevant. In order that the relevant elements of the group may be found, it is necessary to collect them into classes. The elements of a class are related to each other by means of a similarity transformation. If A, B and C

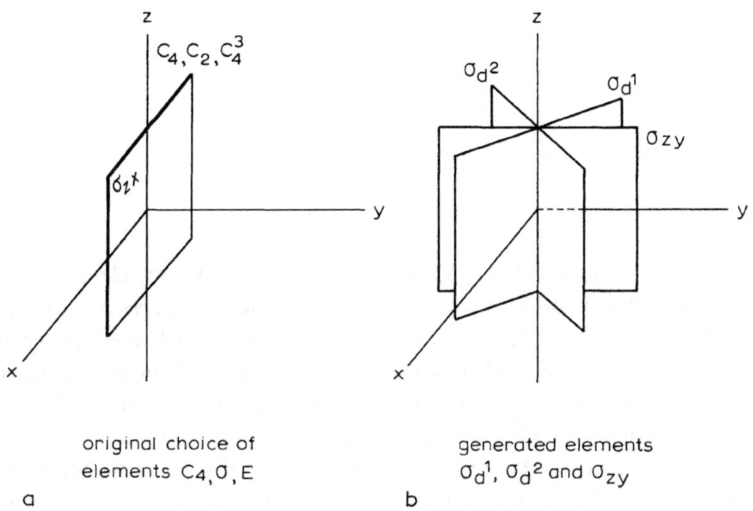

Fig. III-1.

are the elements of such a class, then these transformations are:

$$\begin{aligned} A &= B^{-1}CB = C^{-1}BC \\ B &= A^{-1}CA = C^{-1}AC \\ C &= B^{-1}AB = A^{-1}BB^{-1}. \end{aligned} \qquad \text{(III, 2-1c)}$$

It is easily seen that the following classes may be formed:

$$E; \quad C_4, C_4^3; \quad C_2; \quad \sigma_v(xz), \sigma_v(yz); \quad \sigma_d^1, \sigma_d^2. \qquad \text{(III, 2-1d)}$$

Thus, there are 8 symmetry elements and they are divided into five classes. Three of them contain two elements and are of order two, and the other classes are of order one. It remains to be shown, however, that these elements, taken together in the classes, constitute a group. In order to prove that this is indeed the case, the conditions outlined in the beginning of this paragraph should be investigated. Employing matrix notation for the symmetry elements and matrix multiplication as an operation for combining the elements, it may be shown that the elements constitute a group which is one of the thirty-two point groups. In general, it is found that within point and space groups there are smaller groups. These groups are called subgroups, and the order of the subgroups must always be smaller than the order of the point groups. To be precise, the order of the subgroup is a divisor of the order of the whole group. Sometimes it is found that the subgroup consists of whole classes of elements of the whole group, and these subgroups are called invariant, normal or self-conjugate subgroups. Subgroups play an important role in arriving at the definition of factor groups.

If a group G of order g consists of the elements $E, S_2, S_3, \ldots$, then this group may be written as:

$$G \equiv \{E, S_2, S_3, \ldots S_g\} \quad \text{(III, 2-2)}$$

and the subgroup A consists of the elements $E, S_2, S_3 \ldots$

Thus:

$$A \equiv \{E, S_2, S_3, \ldots S_a\} \quad \text{(III, 2-3)}$$

where $a<g$. The elements $S_{a+1}, S_{a+2} \ldots, S_g$ are obviously not elements of the subgroup A, but this is also the case for the elements $S_{a+1}E, S_{a+1}S_2, \ldots S_{a+1}S_a$; $S_{a+2}E \ldots S_{a+2}S_a$; $S_gE \ldots S_gS_a$. These elements make up the left coset generated by the elements $S_{a+1}, S_{a+2}, \ldots, S_g$ and the elements of the group G. Similarly, the right coset is generated by the product of elements of G and $S_{a+1}, S_{a+2} \ldots$ etc. Both left and right cosets cannot contain the identity element E. The factor group is now defined as the group of the left cosets of the invariant subgroup and thus, if A is such a group, then the elements of the factor group are those mentioned above. Factor groups are important in the analysis of spectra of solids, and the above may be clarified as follows.

Groups which contain the primitive translation as symmetry elements are at first glance groups of infinite order. This can be seen from (III, 1-7). The constants t_x, t_y and t_z can have any integral value. In these infinite groups of primitive translations however, there exist finite groups – denoted by J – and they are the invariant subgroups of the 73 space point groups. The infinite group can be written as:

$$G \equiv \{a_1 t_1 \ldots a_i t_i, E, S_{i+2}, \ldots S_{i+n}\} \quad \text{(III, 2-3)}$$

and the finite group J is associated with the group of elements $a_1 t_1 \ldots a_i t_i$ (where i is large and t_i stands for t_{ix}, t_{iy} and t_{iz}). The factor group, according to the definition given above, contains products of one element of J and the elements $E_1 S_{i+2} \ldots, S_{i+n}$. There is a one-to-one correspondence of the elements of the factor groups and these of the 32 point groups; in other words, the factor group is isomorphic to the corresponding point group.

III-3. Representations of Groups

In section III-2 it was remarked that the matrices representing the symmetry elements E, C_4, C_2, σ_v and σ_d comprise a group. The product of any of these five matrices is a matrix which is related to a symmetry element. The results of all possible products lead to the construction of the group multiplication Table III-2. Such multiplication tables belong to all 32 point groups (and others). They play an important role, as shall become clear later on in this Section. The matrices used to construct the Table III-2 are also given, and any set of matrices which satisfies the group multiplication table forms a representation of the group. The set of matrices of Table III-2 is thus one representation and it is worthwhile to recall here that the set was constructed in studying the transformation properties of a general point in physical space. The question may well be raised whether there are perhaps other ways to find representations.

TABLE III-2

Group multiplication table for the point group C_{4v} and matrices representing the symmetry element E, C_4, C_2, σ_v, σ_d.

Group multiplication table

	E	$2C_4$	C_2	$2\sigma_v$	$2\sigma_d$
E	E	C_4	C_2	σ_v	σ_d
$2C_4$	C_4	E	C_4	σ_d	σ_v
C_2	C_2	C_4	E	σ_v	σ_d
$2\sigma_v$	σ_v	σ_d	σ_v	E	C_4
$2\sigma_d$	σ_d	σ_v	σ_d	C_4	E

Symmetry element	E	C_4^1	C_4^3	C_2
matrix (for a general point in space)	$\begin{pmatrix} 1 & 0 & 0 \\ 0 & 1 & 0 \\ 0 & 0 & 1 \end{pmatrix}$	$\begin{pmatrix} 0 & 1 & 0 \\ -1 & 0 & 0 \\ 0 & 0 & 1 \end{pmatrix}$	$\begin{pmatrix} 0 & -1 & 0 \\ 1 & 0 & 0 \\ 0 & 0 & 1 \end{pmatrix}$	$\begin{pmatrix} -1 & 0 & 0 \\ 0 & -1 & 0 \\ 0 & 0 & 1 \end{pmatrix}$

Symmetry element	$\sigma_v^{(1)}$	$\sigma_v^{(2)}$	$\sigma_d^{(1)}$	$\sigma_d^{(2)}$
matrix (for a general point in space)	$\begin{pmatrix} -1 & 0 & 0 \\ 0 & 1 & 0 \\ 0 & 0 & 1 \end{pmatrix}$	$\begin{pmatrix} 1 & 0 & 0 \\ 0 & -1 & 0 \\ 0 & 0 & 1 \end{pmatrix}$	$\begin{pmatrix} 0 & 1 & 0 \\ 1 & 0 & 0 \\ 0 & 0 & 1 \end{pmatrix}$	$\begin{pmatrix} 0 & -1 & 0 \\ -1 & 0 & 0 \\ 0 & 0 & 1 \end{pmatrix}$

Consider, for example, the unit matrix (1) and let $E \equiv (1)$; $C_4 \equiv (1)$; $C_2 \equiv (1)$; $\sigma_v \equiv (1)$ and $\sigma_d \equiv (1)$. Obviously this set of simple matrices satisfies the group multiplication table and it now appears that the original choice was perhaps not so fortunate, because the set $E \equiv (1)$; $C_4 \equiv (1)$; $C_2 \equiv (1)$; $\sigma_v \equiv (1)$; $\sigma_d(1)$ is also a representation of the group. This may be seen from Table III-3 where a comparison is made between two multiplication tables. The Table III-3A is constructed by substituting the value of $+(1)$ (or $-(1)$) for the appropriate elements generated in Table III-2. The Table III-3B is

TABLE III-3

Group multiplication tables for two sets of representations. The table (A) is obtained by substituting the values $+(1)$ or $-(1)$ in the appropriate entries of the multiplication table given in Table III-2. The results of table (B) are obtained by direct multiplication.

Representation I: $E \equiv (1)$; $C_4 \equiv (1)$; $C_2 \equiv (1)$; $\sigma_v \equiv (1)$; $\sigma_d \equiv (1)$. All are unit matrices.

(A)

	E	$2C_4$	C_2	$2\sigma_v$	$2\sigma_d$
E	+(1)	+(1)	+(1)	+(1)	+(1)
$2C_4$	+(1)	+(1)	+(1)	+(1)	+(1)
C_2	+(1)	+(1)	+(1)	+(1)	+(1)
$2\sigma_v$	+(1)	+(1)	+(1)	+(1)	+(1)
$2\sigma_d$	+(1)	+(1)	+(1)	+(1)	+(1)

(B)

	E	$2C_4$	C_2	$2\sigma_v$	$2\sigma_d$
E	+(1)	+(1)	+(1)	+(1)	+(1)
$2C_4$	+(1)	+(1)	+(1)	+(1)	+(1)
C_2	+(1)	+(1)	+(1)	+(1)	+(1)
$2\sigma_v$	+(1)	+(1)	+(1)	+(1)	+(1)
$2\sigma_d$	+(1)	+(1)	+(1)	+(1)	+(1)

Representation II: $E \equiv (1)$; $C_4 \equiv (1)$; $C_2 \equiv -(1)$; $\sigma_v \equiv -(1)$; $\sigma_d \equiv -(1)$

(A)

	E	$2C_4$	C_2	$2\sigma_v$	$2\sigma_d$
E	+(1)	+(1)	−(1)	−(1)	−(1)
$2C_4$	+(1)	+(1)	+(1)	−(1)	−(1)
C_2	−(1)	+(1)	+(1)	−(1)	−(1)
$2\sigma_v$	−(1)	−(1)	−(1)	+(1)	+(1)
$2\sigma_d$	−(1)	−(1)	−(1)	+(1)	+(1)

(B)

	E	$2C_4$	C_2	$2\sigma_v$	$2\sigma_d$
E	+(1)	+(1)	−(1)	−(1)	−(1)
$2C_4$	+(1)	+(1)	−(1)	−(1)	−(1)
C_2	−(1)	−(1)	+(1)	+(1)	+(1)
$2\sigma_v$	−(1)	−(1)	+(1)	+(1)	+(1)
$2\sigma_d$	−(1)	−(1)	+(1)	+(1)	+(1)

arrived at by actual multiplication. Both tables are identical, but this is not the case for the set of matrices $E \equiv (1)$; $C_4 \equiv (1)$; $C_2 \equiv -(1)$; $\sigma_v \equiv -(1)$ and $\sigma_d \equiv -(1)$. The latter is not a representation of the group because this set clearly does not satisfy the group multiplication table. A search for other combinations of the unit matrices reveals that there are a total of four of these unique sets. The fifth set is made up of matrices of two rows and two columns, and this representation is two-dimensional. The five representations are collected in Table III-4. An example of a set of 3×3 matrices has already been discussed in detail. They are given in Table III-2. However, a close look at these matrices reveals that they can actually be blocked out in 2×2 and 1×1 matrices. If this is done, an interesting fact emerges. A comparison of Table III-2 and Table III-4 shows that the blocked-out matrices are representations of the group.

Situations do arise where sets of 4×4 or higher order matrices are generated, but a study of all these matrices shows that they can always be blocked out in simple matrices which are associated with the five representations of Table III-4. These five sets are thus unique, and they are called the irreducible representations. It is customary to

TABLE III-4

Representations of the group C_{4v}

Elements → / Representation ↓	E	$2C_4$	C_2	$2\sigma_v$	$2\sigma_d$
$\Gamma_1 \equiv A_1$	$+(1)$	$+(1)$	$+(1)$	$+(1)$	$+(1)$
$\Gamma_2 \equiv A_2$	$+(1)$	$+(1)$	$+(1)$	$-(1)$	$-(1)$
$\Gamma_3 \equiv B_1$	$+(1)$	$-(1)$	$+(1)$	$+(1)$	$-(1)$
$\Gamma_4 \equiv B_2$	$+(1)$	$-(1)$	$+(1)$	$-(1)$	$+(1)$
$\Gamma_5 \equiv E$	$\begin{pmatrix}1 & 0\\ 0 & 1\end{pmatrix}$	$\begin{pmatrix}0 & -1\\ +1 & 0\end{pmatrix}$	$\begin{pmatrix}-1 & 0\\ 0 & -1\end{pmatrix}$	$\begin{pmatrix}-1 & 0\\ 0 & +1\end{pmatrix}$	$\begin{pmatrix}0 & 1\\ 1 & 0\end{pmatrix}$
	$\begin{pmatrix}1 & 0\\ 0 & 1\end{pmatrix}$	$\begin{pmatrix}0 & +1\\ -1 & 0\end{pmatrix}$	$\begin{pmatrix}-1 & 0\\ 0 & -1\end{pmatrix}$	$\begin{pmatrix}1 & 0\\ 0 & -1\end{pmatrix}$	$\begin{pmatrix}0 & -1\\ -1 & -0\end{pmatrix}$

The entries are matrices.

represent a group by a table in which the characters of the matrices of the irreducible representations are displayed. The character of a matrix is equal to the sum of the diagonal elements. The character table for the group C_{4v}, having the symmetry elements E, C_4, C_2, σ_v and σ_d is shown in Appendix I, together with such tables for other point groups.

The characters of the matrices of the irreducible representations have some specific properties. They will be discussed here only superficially. Certain symbols shall be used in the discussion, and although some of them have already been defined, it seems appropriate to collect all of them in the Table III-5. The following relations

TABLE III-5

Symbols used to describe Point Groups

E, $C_n{}^m$, i, $S_n{}^m$,	symmetry elements
R	symmetry operation
h	order of the group
k	order of a class
Γ_i	ith irreducible representation
$X^i(R)$	character of the Rth symmetry operation of Γ_i.
l_i	dimension of the matrix R of the irreducible representation
$\Gamma_i(R_{mn})$	the element of the mth row and nth column of the matrix R of T_i

are valid for the matrices of the representations:

$$\sum_R \Gamma_i(R_{mn})\,\Gamma_j(R_{mn}) = 0 \quad \text{if} \quad i \neq j \tag{III, 3-1a}$$

$$\sum_R \Gamma_i(R_{mn})\,\Gamma_i(R_{m'n'}) = 0 \quad \text{if} \quad m = m' \quad \text{and/or} \quad n = n' \tag{III, 3-1b}$$

$$\sum_R \Gamma_i(R_{mn})\,\Gamma_i(R_{mn}) = \frac{h}{l_i}. \tag{III, 3-1c}$$

These three equations can be clarified on the basis that corresponding elements of matrices of one irreducible representation may be taken as the components of a vector in h-dimensional space. The number of vectors is dependent on the dimension of the matrices and the number of irreducible representations. If the vectors are chosen from matrices of different representations then they are orthogonal (III, 3-1a). If the vectors are selected from one particular multi-dimensional representation, then these vectors are also orthogonal (III, 3-1b). The square of the length of any of the vectors is equal to h/l_i, and this result is obtained from (III, 3-1c).

This section is concluded with some remarks on reducible representation. Of particular interest here is the method employed to find the decomposition of a reducible representation into the irreducible representation. Under the symmetry operations of the group, the relation between the original and the transformed physical quantity may be represented by employing matrix notation. (The transformation of a general point in space is one particular example.) If the character of the transformation matrix R is denoted by $\chi(R)$, then the number of times (n_i) the irreducible representation Γ_i occurs is given by:

$$n_i = \frac{1}{h} \sum_i k\chi(R)\, \chi^i(R). \qquad \text{(III, 3-2)}$$

This expression is rather important, and extensive use will be made in the following chapters. The reader may test (III, 3-2) to find the irreducible representations for the specific example of the reducible matrices of Table III-1. (Answer: $\Gamma_{\text{red}} = \Gamma_1 + \Gamma_5$.)

III-4. Transformation Properties of the Scattering Tensor

The apparatus of group theory is most useful in many areas of physics and chemistry, particularly in explaining selection rules of transitions for atoms, ions and molecules. In order to find these selection rules one studies the behaviour of certain functions which are submitted to the symmetry operation of the group which describes the symmetry of the particle.

In Chapter II it was shown that the transformation properties of the element α_{xy} of a tensor α are similar to those of the product xy. Of interest to Raman spectroscopy are the transformation properties of the elements of the scattering tensor. For the purpose of finding the transformation properties, the tensor α may thus be written as:

$$\alpha \equiv x^2 + xy + \ldots + z^2 . \qquad \text{(III, 4-1)}$$

This expression is more complicated than any of the functions considered in the preceding paragraph. Nevertheless, some useful information is obtained if this tensor is subjected to the symmetry operations $\hat{E}$, $\hat{C}_4$, $\hat{C}_2$, $\hat{\sigma}_v$ and $\hat{\sigma}_d$ of the group C_{4v}. It is easily seen that:

$$\alpha \xrightarrow{\hat{E}} \alpha \qquad \text{(III, 4-2a)}$$

and in writing down the complete function, the following matrix is generated:

$$\begin{array}{c|cccccc} & x^2 & xy & xz & y^2 & yz & z^2 \\ \hline x^2 & 1 & 0 & 0 & 0 & 0 & 0 \\ xy & 0 & 1 & 0 & 0 & 0 & 0 \\ xz & 0 & 0 & 1 & 0 & 0 & 0 \\ y^2 & 0 & 0 & 0 & 1 & 0 & 0 \\ yz & 0 & 0 & 0 & 0 & 1 & 0 \\ z^2 & 0 & 0 & 0 & 0 & 0 & 1 \end{array} \qquad \text{(III, 4-2b)}$$

The operation $\hat{E}$ is thus represented by a 6×6 matrix and $\chi(E)=6$. For the symmetry element C_4 the matrix becomes:

$$\begin{array}{c|cccccc} & x^2 & y^2 & z^2 & xy & yz & zx \\ \hline x^2 & 0 & 1 & & & & \\ y^2 & 1 & 0 & & & & \\ z^2 & & & 1 & & & \\ xy & & & & -1 & & \\ yz & & & & & 0 & 1 \\ zx & & & & & 1 & 0 \end{array} \qquad \text{(III, 4-2c)}$$

The character of the transformation matrix is $\chi(C_4)=0$. This routine may now be applied to find the characters of the operations $\hat{C}_2$, $\hat{\sigma}_v$ and $\hat{\sigma}_d$. The values are $\chi(C_2)=2$; $\chi(\sigma_v)=2$; $\chi(\sigma_d)=2$. The next step is to see if this representation of a set of 6×6 matrices is reducible via (III, 3-2). This is not the case, and the function (III, 4-1) is not a proper *basic function* for the irreducible representations of the group. The function z^2 is, however, a proper basis function because the characters are:

$$\chi(E)=1; \quad \chi(C_4)=1; \quad \chi(C_2)=1; \quad \chi(\sigma_v)=1 \quad \text{and} \quad \chi(\sigma_d)=1\,.$$

The function z^2 belongs to the irreducible representation A_1 of the group C_{4v}, but the functions x^2 and y^2 are again found to be improper basis functions. The matrices describing the transformation of the combination x^2+y^2 are all of two rows and two columns. The characters are $\chi(E)=2$; $\chi(C_4)=0$; $\chi(C_2)=2$; $\chi(\sigma_v=2)$; $\chi(\sigma_d)=0$. From the character table of the group C_{4v} and Equation (III, 3-2), it follows that the number of times the totally symmetric representation occurs is:

$$n_{A_1} = \tfrac{1}{8}\,[1.2.1 + 2.0.1 + 1.2.1 + 2.2.1 + 2.0.2] = 1\,. \qquad \text{(III, 4-3)}$$

From the above it is concluded that the components α_{zz} and $\alpha_{xx}+\alpha_{yy}$ transform similar to the totally symmetric representation of the group. As a direct consequence of the presence of a four-fold rotation axis, the coordinate x is carried over into the coordinate y and vice versa. These coordinates transform similar to the irreducible representation E and they are inseparable. This is indicated in the ninth column of the table for C_{4v} by placing a comma between x and y. The components x^2 and y^2 are also inseparable, and thus $\alpha_{xx}=\alpha_{yy}$; but the component α_{z^2} is different from α_{xx} or α_{yy}. The notation follwed here is: $\alpha_{xx}+\alpha_{yy}$, α_{zz} and the two tensors which transform according to the totally symmetric representation are:

$$\begin{pmatrix} a & 0 & 0 \\ 0 & a & 0 \\ 0 & 0 & 0 \end{pmatrix}; \quad \begin{pmatrix} 0 & 0 & 0 \\ 0 & 0 & 0 \\ 0 & 0 & b \end{pmatrix} \qquad \text{(III, 4-4)}$$

where

$$\alpha_{xx} = \alpha_{yy} = a \quad \text{and} \quad \alpha_{zz} = b\,.$$

The expression (III, 4-1) for the tensor α is restrictive in that only a combination of positive contributions is chosen. For instance, it can be shown that the combination x^2-y^2 is a proper basis function for the representation B_1. The tensor associated with this combination is:

$$\begin{pmatrix} c & 0 & 0 \\ 0 & -c & 0 \\ 0 & 0 & 0 \end{pmatrix} \qquad \text{(III, 4-5)}$$

where

$$c = \alpha_{xx} = \alpha_{yy}$$

The components α_{xz} and α_{yz} transforms according to E and they are indistinguishable. However, if the function xz transforms according to E, then zx also transforms like E, as well as do $xz+zx$. Consequently, the tensors which belong to E are:

$$\begin{pmatrix} 0 & 0 & d \\ 0 & 0 & 0 \\ d & 0 & 0 \end{pmatrix}; \quad \begin{pmatrix} 0 & 0 & 0 \\ 0 & 0 & d \\ 0 & d & 0 \end{pmatrix} \qquad \text{(III, 4-6)}$$

The component α_{xy} and thus also α_{yx} transforms like the irreducible representation (species) B_3 of C_{4v}.

The tensors considered here are components or combinations of components of a symmetric tensor and there rests the problem of finding the transformation properties of the antisymmetric tensors $\alpha_{xy}-\alpha_{yx}$; $\alpha_{xz}-\alpha_{zx}$; $\alpha_{yz}-\alpha_{zy}$; or, in matrix notation:

$$\begin{pmatrix} 0 & \alpha_{xy} & 0 \\ -\alpha_{yx} & 0 & 0 \\ 0 & 0 & 0 \end{pmatrix}; \quad \begin{pmatrix} 0 & 0 & \alpha_{xz} \\ 0 & 0 & 0 \\ -\alpha_{zx} & 0 & 0 \end{pmatrix}; \quad \begin{pmatrix} 0 & 0 & 0 \\ 0 & 0 & \alpha_{yz} \\ 0 & -\alpha_{zy} & 0 \end{pmatrix} \qquad \text{(III, 4-7)}$$

According to the discussion of vectors and tensors offered in Chapter II, the tensors (III, 4-7) can be represented by the cross product of ordinary vectors. These cross products are axial vectors and the transformation properties of the tensor

$$\begin{pmatrix} 0 & \alpha_{xy} & 0 \\ -\alpha_{yx} & 0 & 0 \\ 0 & 0 & 0 \end{pmatrix}$$

are thus similar to those of the vector $\mathbf{R}_z=\mathbf{x}\times\mathbf{y}$, where $\mathbf{R}_z$ is perpendicular to the xy plane. This axial vector changes sign upon reflection through $\sigma(zx)$ and $\sigma(zy)$, but

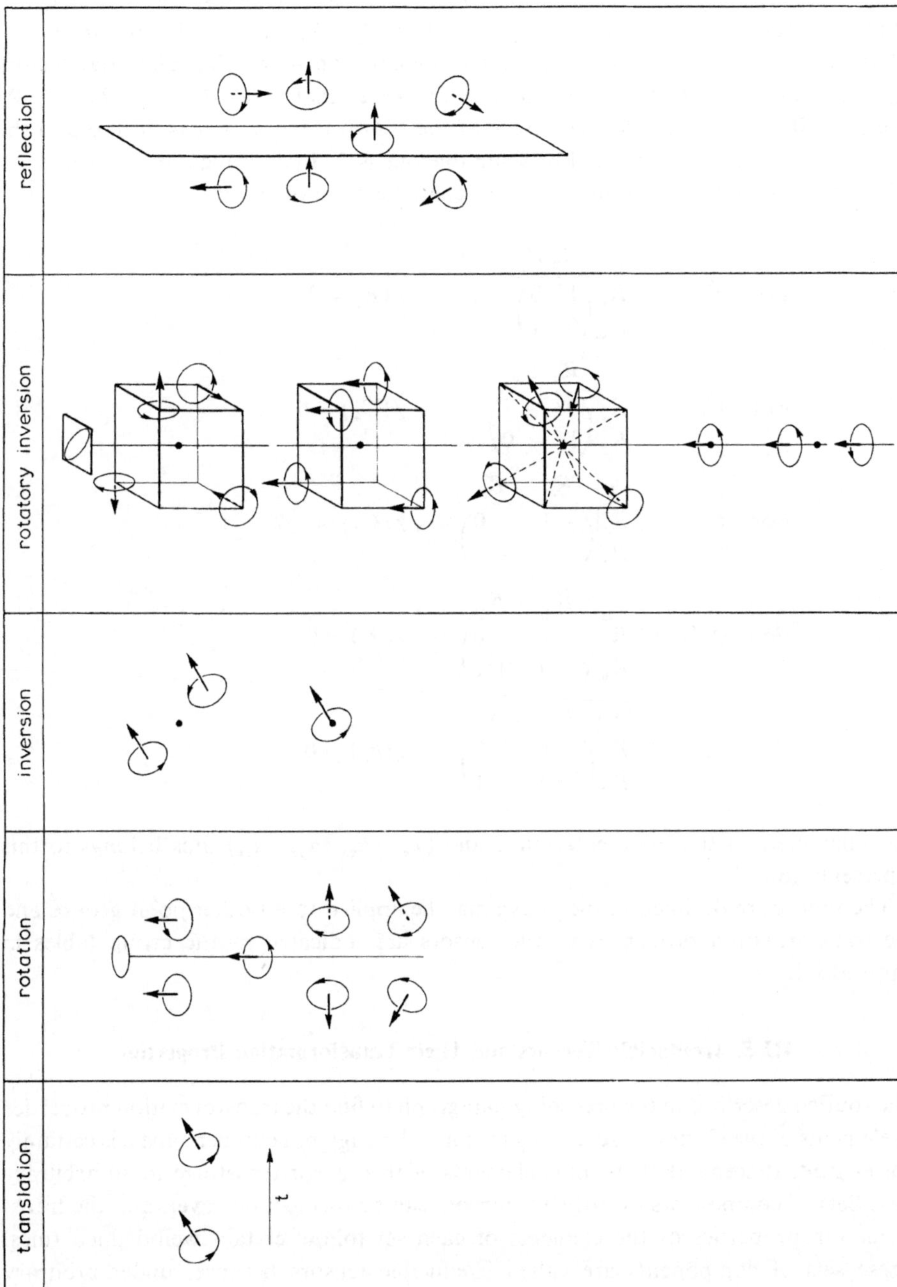

Fig. III-2. Symmetry operations on axial vectors.

goes over into itself under the operations $\hat{E}, \hat{C}_4$ and $\hat{C}_2$. In Figure III-2 is shown the behaviour of the axial vector for some of the symmetry elements. The characters of of the transformations are $\chi(E)=1$; $\chi(C_4)=1$; $\chi(C_2)=1$; $\chi(\sigma_v)=-1$ and $\chi(\sigma_d)=-1$. Thus, $\mathbf{R}_z$ transforms like the irreducible representation A_2 of C_{4v}. The axial vectors $\mathbf{R}_x$ and $\mathbf{R}_y$ do not form a basis function, but it is seen that under the operation $\hat{C}_4$, $\mathbf{R}_x$ goes into $\mathbf{R}_y$ and $\mathbf{R}_y$ into $\mathbf{R}_x$. The trace of the transformation matrix is zero and the only representation which could be a suitable one is E. In carrying out the operations of the group, the following matrices and characters are obtained:

For $\hat{E}$
$$\begin{array}{c|cc} & R_x & R_y \\ \hline R_x & 1 & 0 \\ R_y & 0 & 1 \end{array} \qquad \chi(\hat{E}) = 2.$$

For $\hat{C}_4$
$$\begin{array}{c|cc} & R_x & R_y \\ \hline R_x & 0 & -1 \\ R_y & 1 & 0 \end{array} \qquad \chi(\hat{C}_4) = 0$$

For $\hat{C}_2$
$$\begin{array}{c|cc} & R_x & R_y \\ \hline R_x & -1 & 0 \\ R_y & 0 & -1 \end{array} \qquad \chi(\hat{C}_2) = -2$$

For $\hat{\sigma}_v(xz)$
$$\begin{array}{c|cc} & R_x & R_y \\ \hline R_x & -1 & 0 \\ R_y & 0 & +1 \end{array} \qquad \chi(\hat{\sigma}_v) = 0$$

For $\hat{\sigma}_d$
$$\begin{array}{c|cc} R & R_x & R_y \\ \hline R_x & -1 & 0 \\ R_y & 0 & +1 \end{array} \qquad \chi(\hat{\sigma}_d) = 0$$

The characters fit the representation E and $(\alpha_{xz}-\alpha_{zx}, \alpha_{yz}-\alpha_{zy})$ thus belongs to this representation.

The procedure outlined in the above may be applied to all other point groups and the transformation properties of the tensors are indicated in the group tables in Appendix I.

III-5. Irreducible Tensors and Their Transformation Properties

The routine described in the preceding paragraph to find the transformation properties of elements of the Cartesian scattering tensor – although not cumbersome – is certainly not elegant. It seems that the nine elements of this tensor transform in an arbitrary way. Sets of components of ordinary tensors can be formed, however, and the transformation properties of the elements of each set follow certain well-defined rules. These sets of components are called irreducible tensors because, under arbitrary rotations, the members of a set transform only into combinations of members of the same set. In order to arrive at the irreducible tensor sets, detailed use has to be made of the mathematics which is encountered in atomic physics. These details which relate to commutation properties of the angular momentun or its components, and the coupling of angular momenta fall outside the scope of this book and, where necessary, only the results of properties are used. The irreducible tensors are not Cartesian tensors. Instead, use is made of circular coordinates. An element of the Cartesian scattering tensor for the transition $k \leftrightarrow n$ is given by

$$(\alpha_{\varrho\sigma})_{kn} = \frac{1}{h} \sum_r \left[\frac{(M_\sigma)_{kr}\,(M_\varrho)_{rn}}{\nu_{rk} - \nu_0} + \frac{[\varrho \leftrightarrow \sigma]}{\nu_{rn} + \nu_0} \right]. \tag{III, 5-1}$$

The moment $(M_\sigma)_{kr}$ is the transition moment of a component of the electric dipole operator along the axis of a Cartesian coordinate system. If $\mathbf{r}$ is a position vector, then such moments are generally given by the integral $\int \psi_r^* e\mathbf{r}\psi_k \, d\tau$. The circular components of the position vector are defined by:

$$r_0 = z$$

$$r_{\pm 1} = \pm \frac{1}{\sqrt{2}} (x \mp iy). \tag{III, 5-2}$$

Moments A and B can now be formed using the circular components so that

$$A_\lambda = \int \Psi_r^* r_\lambda \Psi_k \, d\tau = (M_{r_\lambda})_{kr}$$

and

$$B_\mu = \int \Psi_n^* r_\mu \Psi_r \, d\tau = (M_{r_\mu})_{rn} \tag{III, 5-3}$$

where λ and μ are equal to 0 and/or ± 1. Analogous to (III, 5-1), the expression for an element $(\alpha_{\lambda\mu})_{kn}$ of the reducible spherical tensor is then given by:

$$(\alpha_{\lambda\mu})_{kn} = \frac{1}{h} \sum_r \left[\frac{A_\lambda B_\mu}{\nu_{rk} - \nu_0} + \frac{[\lambda \leftrightarrow \mu]}{\nu_{rn} + \nu_0} \right]. \tag{III, 5-4a}$$

The nine elements of (III, 5-4) in terms of the elements of the Cartesian scattering tensor are obtained from (III, 5-4) and (III, 5-3). They are:

$$\alpha_{00} = A_0 B_0 = \alpha_{zz}$$

$$\alpha_{11} = A_{+1} B_{+1} = \tfrac{1}{2}(\alpha_{xx} - \alpha_{yy} + i(\alpha_{xy} + \alpha_{yx}))$$

$$\alpha_{01} = A_0 B_{+1} = -\frac{1}{\sqrt{2}} (\alpha_{zx} + i\alpha_{zy})$$

$$\alpha_{1-1} = A_{+1} B_{-1} = -\tfrac{1}{2}(\alpha_{xx} + \alpha_{yy} + i(\alpha_{yx} - \alpha_{xy}))$$

$$\alpha_{10} = A_{+1} B_0 = -\frac{1}{\sqrt{2}} (\alpha_{xz} + i\alpha_{yz}) \tag{III, 5-4b}$$

$$\alpha_{-11} = A_{-1}B_{+1} = -\tfrac{1}{2}\left(\alpha_{xx} + \alpha_{yy} + i\left(\alpha_{xy} - \alpha_{yx}\right)\right)$$

$$\alpha_{0-1} = A_0 B_{-1} = \frac{1}{\sqrt{2}}\left(\alpha_{xz} - i\alpha_{yz}\right)$$

$$\alpha_{-1-1} = A_{-1}B_{-1} = \tfrac{1}{2}\left(\alpha_{xx} - \alpha_{yy} - i\left(\alpha_{xy} + \alpha_{yx}\right)\right)$$

$$\alpha_{-10} = A_{-1}B_0 = \frac{1}{\sqrt{2}}\left(\alpha_{zx} - i\alpha_{zy}\right). \qquad \text{(III, 5-4b)}$$

The transition moments A_λ, B_μ may be looked upon as ordinary or polar vectors. Such vectors behave under arbitrary rotations like axial vectors; an example of such an axial vector is the angular momentum J. Coupling of angular momenta is subjected to specific rules. If the mathematical apparatus* is applied to the problem of finding the components of the irreducible tensors α_Q^K, in relation to the components of the reducible spherical tensor $\alpha_{y\mu}$, then this relation is given by:

$$\alpha_Q^K = \sum_{\lambda\mu} A_\lambda B_\mu C_{\lambda\mu KQ} = \sum_{\lambda\mu} \alpha_{\lambda\mu} C_{\lambda\mu KQ}. \qquad \text{(III, 5-5)}$$

The constants $C_{\lambda\mu kQ}$ are known as Clebsch-Gordan coefficients. These coefficients may be calculated, and the connection of the components α_Q^K and $\alpha_{\varrho\sigma}$ is given in the following Table III-6. The transformation is unitary and orthogonal. Thus an element α_{xx} expressed in the α_Q^K's is given by:

$$\alpha_{xx} = -\frac{1}{\sqrt{3}}\alpha_0^0 + \tfrac{1}{2}\alpha_2^2 - \frac{1}{\sqrt{6}}\alpha_0^2 + \tfrac{1}{2}\alpha_{-2}^2 \qquad \text{(III, 5-6a)}$$

but on the other hand:

$$\alpha_0^0 = -\frac{1}{\sqrt{3}}\left(\alpha_{xx} + \alpha_{yy} + \alpha_{zz}\right)$$

$$\left.\begin{aligned}
\alpha_1^1 &= \tfrac{1}{2}\left(\alpha_{zx} - \alpha_{xz} - i\alpha_{yz} + i\alpha_{zy}\right)\\
\alpha_0^1 &= \frac{1}{\sqrt{2}}\left(i\alpha_{xy} - i\alpha_{yx}\right)\\
\alpha_{-1}^1 &= -\tfrac{1}{2}\left(\alpha_{xz} - \alpha_{zx} - i\alpha_{yz} + i\alpha_{zy}\right)
\end{aligned}\right\}$$

$$\left.\begin{aligned}
\alpha_2^2 &= -\tfrac{1}{2}\left(\alpha_{xx} - \alpha_{yy} + i\alpha_{xy} + i\alpha_{yx}\right)\\
\alpha_1^2 &= -\tfrac{1}{2}\left(\alpha_{xz} + \alpha_{zx} + i\alpha_{yz} + i\alpha_{zy}\right)\\
\alpha_0^2 &= \frac{1}{\sqrt{6}}\left(2\alpha_{zz} - \alpha_{xx} - \alpha_{yy}\right)\\
\alpha_{-1}^2 &= \tfrac{1}{2}\left(\alpha_{xz} + \alpha_{zx} - i\alpha_{yz} - i\alpha_{zy}\right)\\
\alpha_{-2}^2 &= \tfrac{1}{2}\left(\alpha_{xx} - \alpha_{yy} - i\alpha_{xy} - i\alpha_{yx}\right).
\end{aligned}\right\} \qquad \text{(III, 5-6b)}$$

The irreducible tensors are: α_0^0, $\alpha_Q^1(Q = -1, 0, 1)$ and $\alpha_Q^2(Q = 2, 1, 0, -1, -2)$. Of course such tensors are not restricted to the scattering tensors. They may be derived

* See also Chapter IV-7, p. 122.

TABLE III-6

The unitary matric (u) connecting the castesion and irreducible tensor components

	$U_j \downarrow$ α_{xx}	α_{yy}	α_{zz}	α_{xz}	α_{zx}	α_{yz}	α_{zy}	α_{xy}	α_{yx}
$U_i \rightarrow$ $\alpha_0{}^0$	$-1/\sqrt{3}$	$-1/\sqrt{3}$	$-1/\sqrt{3}$	–	–	–	–	–	–
$\alpha_1{}^1$	–	–	–	$-1/2$	$+1/2$	$-i/2$	$+i/2$	–	–
$\alpha_0{}^1$	–	–	–	–	–	–	–	$i/\sqrt{2}$	$-i/\sqrt{2}$
$\alpha_{-1}{}^1$	–	–	–	$-1/2$	$+1/2$	$+i/2$	$-i/2$	–	–
$\alpha_2{}^2$	$+1/2$	$-1/2$	–	–	–	–	–	$i/2$	$i/2$
$\alpha_1{}^2$	–	–	–	$-1/2$	$-1/2$	$-i/2$	$-i/2$	–	–
$\alpha_0{}^2$	$-1/\sqrt{6}$	$-1/\sqrt{6}$	$+2/\sqrt{6}$	–	–	–	–	–	–
$\alpha_{-1}{}^2$	–	–	–	$+1/2$	$+1/2$	$-i/2$	$-i/2$	–	–
$\alpha_{-2}{}^2$	$+1/2$	$-1/2$	–	–	–	–	–	$-i/2$	$-i/2$

$\Sigma\, U_i\, U_j{}^* = 1 \qquad (U_{ji})^{-1} = U_{ji}{}^*$

from the coupling of any two vectors, assuming that such vectors are related to circular coordinates. A common example of irreducible tensors is the set of spherical harmonics $Y^l_m(\theta, \varphi)$, because they also transform among themselves under rotations. This can be shown in the following way. The spherical harmonics can be written as a product of functions which depend on the angles θ and φ as follows:

$$Y^l_m(\theta, \varphi) = \theta^l_m(\theta)\, \phi_m(\varphi) \qquad \text{(III, 5-7a)}$$

where

$$\theta^l_m(\theta) = \sqrt{\frac{2l+1}{2}\frac{(l-|m|)!}{(l+|m|)!}} \cdot P^l_{|m|}(\cos\theta) \qquad \text{(III, 5-7b)}$$

and

$$\phi_m(\varphi) = \frac{1}{\sqrt{2\pi}}\, e^{im\varphi} \quad \text{and} \quad m = -l, -l+1, \ldots l-1, l. \qquad \text{(III, 5-7c)}$$

The functions $\theta^l_m(\theta)$ are the associated Legendre functions and they depend on the angle θ. Some normalized spherical harmonics are given in Table III-7. The table includes the functions for both positive and negative values of m. It is easily seen that:

$$Y^l_{-m}(\theta, \varphi) = (-1)^m (Y^l_m(\theta, \varphi))^* \qquad \text{(III, 5-7d)}$$

where $(Y^l_m)^*$ is the complex conjungate of Y^l_m. Indicated in the table also are relations of the Y^l_m's in terms of the coordinates x, y and z. It is seen that one form in the table is calculated from the other by employing the relations

$$x \pm iy = r \sin\theta\, e^{\pm i\varphi}$$

and

$$z = r\cos\theta\,. \qquad \text{(III, 5-8)}$$

In order to find out if these spherical harmonics are proper basis functions for the group C_{4v}, they have to be submitted to the operations of the group. It is noted that

TABLE III-7

Some normalized spherical harmonics

$l=0$	$Y_0^0 = \dfrac{1}{2\sqrt{\pi}}$
	$Y_{-1}^1 = \dfrac{1}{2}\sqrt{\dfrac{3}{2\pi}}\,\dfrac{x-iy}{r} = \dfrac{1}{2}\sqrt{\dfrac{3}{2\pi}}\sin\theta\, e^{-i\varphi}$
$l=1$	$Y_0^1 = \dfrac{1}{2}\sqrt{\dfrac{3}{\pi}}\,\dfrac{z}{r} = \dfrac{1}{2}\sqrt{\dfrac{3}{\pi}}\cos\theta$
	$Y_1^1 = -\dfrac{1}{2}\sqrt{\dfrac{3}{2\pi}}\,\dfrac{x+iy}{r} = -\dfrac{1}{2}\sqrt{\dfrac{3}{2\pi}}\sin\theta\, e^{i\varphi}$
	$Y_{-2}^2 = \dfrac{1}{4}\sqrt{\dfrac{15}{2\pi}}\,\dfrac{(x-iy)^2}{r^2} = \dfrac{1}{4}\sqrt{\dfrac{15}{2\pi}}\sin^2\theta\, e^{-2i\varphi}$
	$Y_{-1}^2 = \dfrac{1}{2}\sqrt{\dfrac{15}{2\pi}}\,\dfrac{z(x-iy)}{r^2} = \dfrac{1}{2}\sqrt{\dfrac{15}{2\pi}}\sin\theta\cos\theta\, e^{-i\varphi}$
$l=2$	$Y_0^2 = \dfrac{1}{4}\sqrt{\dfrac{5}{\pi}}\,\dfrac{3z^2-r^2}{r^2} = \dfrac{1}{4}\sqrt{\dfrac{5}{\pi}}(3\cos^2\theta-1)$
	$Y_1^2 = -\dfrac{1}{2}\sqrt{\dfrac{15}{2\pi}}\,\dfrac{z(x+iy)}{r^2} = -\dfrac{1}{2}\sqrt{\dfrac{15}{2\pi}}\sin\theta\cos\theta\, e^{i\varphi}$
	$Y_2^2 = \dfrac{1}{4}\sqrt{\dfrac{15}{2\pi}}\,\dfrac{(x+iy)^2}{r^2} = \dfrac{1}{4}\sqrt{\dfrac{15}{2\pi}}\sin^2\theta\, e^{2i\varphi}$

the polar coordinate r and the z-axis- are invariant under operations $\hat{E}$, $\hat{C}_4$, $\hat{C}_2$, $\hat{\sigma}_v$ and $\hat{\sigma}_d$ of the group. Thus the spherical harmonics Y_0^0, Y_0^1 and Y_0^2 (each, separately) are all invariant under the operations of the group. As a direct result, they belong to ȷhe representation A_1 of C_{4v}. In order to find the behaviour of the other spherical harmonics, it seems appropriate to investigate the behaviour of the functions $f^+=x+iy$ and $f^-=x-iy$. By inspection it is found that:

$$\begin{aligned} \hat{C}_4 f^- &= \hat{C}_4(x-iy) = -y-ix = -(y+ix) \\ &= -\left(\frac{i^2}{-1}\right)(y+ix) \\ &= -(x-iy) = -if^- \end{aligned} \qquad \text{(III, 5-9)}$$

and thus

	$\hat{E}$	$\hat{C}_4$	$\hat{C}_2$	$\hat{\sigma}_v$	$\hat{\sigma}_d$
$f^- = x-iy$	f^-	$-if^-$	$-f^-$	$-f^+$	$-if^+$
$f^+ = x+iy$	f^+	if^+	$-f^+$	$-f^-$	if^-

It is noted that f^- and f^+ transform into each other and thus they belong to the two-dimensional representation E. Therefore the combination $(Y_{-1}^1 + Y_{+1}^1)$ transforms like E. The behaviour of the spherical harmonics under the operations of the group C_{4v} may also be given by the following Table – Table III-8

TABLE III-8

Symmetry properties of some spherical Harmonics under operations of the point group C_{4v}

operation → / function ↓	$\hat{E}$	$\hat{C}_4$	$\hat{C}_2$	$\hat{\sigma}_v$	σ_d
$Y_0{}^0$	$Y_0{}^0$	$Y_0{}^0$	$Y_0{}^0$	$Y_0{}^0$	$Y_0{}^0$
$Y_{-1}{}^1$	$Y_{-1}{}^1$	$-iY_1{}^{-1}$	$-Y_{-1}{}^1$	$Y_1{}^1$	$iY_1{}^1$
$Y_0{}^1$	$Y_0{}^1$	$Y_0{}^1$	$Y_0{}^1$	$Y_0{}^1$	$Y_0{}^1$
$Y_1{}^1$	$Y_1{}^1$	$iY_1{}^1$	$-Y_1{}^1$	$Y_1{}^{-1}$	$-iY_{-1}{}^1$
$Y_2{}^2$	$Y_2{}^2$	$-Y_2{}^2$	$Y_2{}^2$	$Y_2{}^{-2}$	$Y_{-2}{}^2$
$Y_1{}^2$	$Y_1{}^2$	$iY_1{}^2$	$-Y_1{}^2$	$Y_2{}^{-1}$	$-iY_{-2}{}^1$
$Y_0{}^2$	$Y_0{}^2$	$Y_0{}^2$	$Y_0{}^2$	$Y_0{}^2$	$Y_0{}^2$
$Y_{-1}{}^2$	$Y_1{}^2$	$-iY_1{}^2$	$-Y_{-1}{}^2$	$Y_2{}^1$	$iY_1{}^2$
$Y_{-2}{}^2$	$Y_{-2}{}^2$	$-Y_{-2}{}^2$	$Y_{-2}{}^2$	$Y_2{}^2$	$-Y_2{}^2$

The table demonstrates in a clear way that the spherical harmonics of one set transform into each other. From the behaviour under the operations of the group, the following are the proper basis functions:

(i) $Y_2^2 + Y_{-2}^2$ transforms like B_1

(ii) $Y_2^2 - Y_{-2}^2$ transforms like B_2

(iii) (Y_1^2, Y_{-1}^2) transforms like E.

The behaviour of the irreducible tensor components α_Q^K under the symmetry operations of point groups is identical to those of the Y_m^l. The basis for the transformation properties in both cases are the spherical coordinates $x \pm iy$ and z. The following equalities hold: $K \equiv l$ and $Q \equiv m$ and the symmetry properties of the α_Q^K's are given in the tenth row of the group tables – see Appendix I. The spherical harmonics play an important role in the transformation properties of atomic orbitals. The wave functions for atomic orbitals are given by:

$$\Psi_{n,l,m} = R_{nl}(r)\, \theta_l^m(\theta)\, \phi_m(\varphi) \qquad \text{(III, 5-10)}$$

The radial function $R_{nl}(r)$ is not dependent on the polar coordinates θ and φ; thus $R_{nl}(r)$ is always invariant to all operations of the group. From the behaviour of the spherical harmonics it follows that α_0^0 transforms like an s-orbital, α_Q^1 transforms like the three p orbitals and the five components of α_Q^2 have symmetry properties equal to these of the five d orbitals. There is, however, a slight difference between the three α_Q^1's and the p_x, p_y and p_z orbitals. The latter change sign under inversion of the coordinate system, but the α_Q^K's behave as the components of an axial vector and they are invariant under this operation.

III-6. The Scattering Operator

From Quantum Mechanics it follows that the average value P of an operator $(\hat{P})$ is given by: $P=\int\psi_n^* \hat{P}\, \psi_k\, d\tau$ or using Diracs notation;

$$P = \langle\psi_n|\hat{P}|\psi_k\rangle . \qquad \text{(III, 6-1)}$$

The function ψ_n^*, ψ_k is used as a weight or probability function, and a further meaning of (III, 6-1) is that the operator $\hat{P}$ is averaged over all configurations in space. A well-known operator is the electric dipole operator $\sum_i e_i\mathbf{r}_i$. An electric dipole transition between the states k and n is allowed if the average value $\sum_i e_i r_i = \langle\psi_n|\sum_i e_i \mathbf{r}_i|\psi_k\rangle$ is different from zero. Another and less known operator is the quadrupole operator $\sum_i e_i \mathbf{r}_i \mathbf{r}_j$. Electric dipole and quadrupole transitions are genuine physical quantities. Such quantities are of course totally independent of the coordinate system of the atom, ion or molecule to which the electrons of charge e_i and position vector $\mathbf{r}_i$ belong. Consequently, the average values of $\sum_i e_i r_i$ and $\sum_i e_i r_i r_j$ are thus functions which are invariant if subjected to the symmetry operations of the group which describes the symmetry of the atom, molecule or ion. From the character tables of the point groups it follows that these physical quantities $\sum_i e_i r_i$ and $\sum_i e_i r_i r_j$ necessarily transform like the totally symmetric representations. Analogous to dipole and quadrupole operators is the scattering operator $\hat{\alpha}_{\varrho\sigma}$ for Rayleigh and Raman processes. The operator is defined as follows. The expression for an element of the scattering tensor $(\alpha_{\varrho\sigma})_{kn}$ for a transition which brings the scattering entity from state k into state n is given by:

$$(\alpha_{\varrho\sigma})_{kn} = \frac{1}{h}\sum_r \left[\frac{\langle\psi_k|\sum_i e_i\sigma|\,\psi_r\rangle\,\langle\psi_r|\sum_i e_i\varrho|\,\psi_n\rangle}{\nu_{rk}-\nu_0} + \frac{[\varrho\leftrightarrow\sigma]}{\nu_{rn}+\nu_0}\right]. \qquad \text{(III, 6-2)}$$

The average value of the scattering operator is defined as:

$$(\alpha_{\varrho\sigma})_{kn} = \langle\psi_n|\hat{\alpha}_{\varrho\sigma}|\,\psi_k\rangle \qquad \text{(III, 6-3a)}$$

where the operator $\hat{\alpha}_{\varrho\sigma}$ is:

$$\hat{\alpha}_{\varrho\sigma} = \frac{1}{h}\sum_r \left[\frac{|\sum_i e_i\varrho\,|\psi_r\rangle\,\langle\psi_r|\,\sum_i e_i\sigma|}{\nu_{rk}-\nu_0} + \frac{[\varrho\leftrightarrow\sigma]}{\nu_{rn}+\nu_0}\right]. \qquad \text{(III, 6-3b)}$$

Thus, in order that a Rayleigh or Raman transition becomes a physically observable quantity, it is a prerequisite that $(\alpha_{\varrho\sigma})_{kn}\neq 0$ and a knowledge of the conditions for which the integrand $\langle\psi_n|\hat{\alpha}_{\varrho\sigma}|\psi_k\rangle\neq 0$ is of paramount importance. It is not necessary to perform a detailed calculation of (III, 6-2). The conditions of nonvanishing of $(\alpha_{\varrho\sigma})_{kn}$ follow directly from the symmetry properties of the wave functions ψ_n^* and ψ_k and the scattering operator $\hat{\alpha}_{\varrho\sigma}$. It is here where some knowledge of group theory pays off. The symmetry properties of $\alpha_{\varrho\sigma}$ were discussed in Sections 4 and 5 of this chapter. Wave functions may also be constructed in such a way that they form a basis for representations of the point groups. The physically observable $(\alpha_{\varrho\sigma})_{kn}$ is thus represented by the triple product of representations:

$$\Gamma_{\psi_n^*} \times \Gamma_{\hat{\alpha}_{\varrho\sigma}} \times \Gamma_{\psi_k} \qquad \text{(III, 6-4)}$$

and the totally symmetric representation of any of the 32 point groups should be contained in this triple product so that $(\alpha_{\varrho\sigma})_{kn} \neq 0$.

The rules of multiplication of representations are defined in that products are taken of the characters of the symmetry elements; and the characters of the representation of a direct product are equal to the products of the characters of the individual representations. If for instance the representations are those of the point group C_{4v} then from the character tables it follows that

$$A_1 \times A_2 = A_2$$
$$A_2 \times A_2 = A_1$$

but also

$$A_2 \times B_1 = B_2$$
$$E \times E = A_1 + A_2 + B_1 + B_2$$

and

$$A_2 \times E \times B_2 = B_1 \times E = E\,. \qquad \text{(III, 6-5)}$$

The above rules can be used to show that the transformation properties of the operator $\hat{\alpha}_{\varrho\sigma}$ and the tensor elements $\alpha_{\varrho\sigma}$ are identical. From Equation (III, 6-2) it follows that the integral

$$\langle\psi_n|\, e\varrho\, |\psi_r\rangle \neq 0 \quad \text{if} \quad \Gamma_{\psi_n^*} \times \Gamma_\varrho \times \Gamma_{\psi_r} \equiv A \qquad \text{(III, 6-6a)}$$

where A is the totally symmetric species of any of the 32 point groups. Thus also

$$\Gamma_{\psi_r} = \Gamma_{\psi_n^*} \times \Gamma_\varrho\,. \qquad \text{(III, 6-6b)}$$

The matrix element

$$\langle\psi_k|\, e\hat{\sigma}\, |\psi_r\rangle \neq 0 \quad \text{if} \quad \Gamma_{\psi_k^*} \times \Gamma_\sigma \times \Gamma_{\psi_r} \equiv A \qquad \text{(III, 6-7)}$$

and a combination of (III, 6-6a) and (III, 6-7) yields:

$$\Gamma_{\psi_k} \times \Gamma_\varrho \times \Gamma_\sigma \times \Gamma_{\psi_n^*} \equiv A \qquad \text{(III, 6-8a)}$$

or

$$\Gamma_{\psi_k} \times \Gamma_{\varrho\sigma} \times \Gamma_{\psi_n^*} \equiv A \qquad \text{(III, 6-8b)}$$

thus

$$\Gamma_{\varrho\sigma} = \Gamma_{\psi_n^*} \times \Gamma_{\psi_k}\,. \qquad \text{(III, 6-8c)}$$

On the other hand the integral

$$\langle\psi_n|\, \hat{\alpha}_{\varrho\sigma}\, |\psi_k\rangle \neq 0 \qquad \text{(III, 6-9a)}$$

if

$$\Gamma_{\psi_n^*} \times \Gamma_{\hat{\alpha}_{\varrho\sigma}} \times \Gamma_{\psi_k} \equiv A \qquad \text{(III, 6-9b)}$$

and from Equation (III, 6-8c) it follows that

$$\Gamma_{\hat{\alpha}_{\varrho\sigma}} \times \Gamma_{\varrho\sigma} \equiv A \qquad \text{(III, 6-10a)}$$

which is true if

$$\Gamma_{\hat{\alpha}_{\varrho\sigma}} \equiv \Gamma_{\varrho\sigma}.$$

Thus the tensor operator $\hat{\alpha}$ transforms like the species A_1, A_2, B_1, B_2 and E of C_{4v}. An inspection of the character table shows that the product of all one-dimensional representations equals one of these representations, and if the wave functions ψ_k and ψ_n belong to any of these representations, then there is always an element of $\hat{\alpha}$ which transforms according to the species appearing in the product and thus $\Gamma_{\psi_n^*} \times \Gamma_{\hat{\alpha}} \times \Gamma_{\psi_k} \equiv A$. If, however, one or both wave functions transform like E, then the totally symmetric representation is *among* the various species which occur in the product of the representations. The selection rules for the Raman transitions become particularly simple if the state in which the transition originates transforms like the totally symmetric representation of the point group. This is the rule rather than the exception. The transition is Raman active if:

$$\Gamma_{\psi_n} = \Gamma_{\hat{\alpha}}.$$

Throughout this chapter the group C_{4v} is employed to clarify matters, and it goes without saying that other point groups could also have been taken. The results are then different but the principle remain the same.

Suggestions for Further Reading

Brandmüller, J. and Moser, H.: 1963, *Einführung in die Ramanspektroskopie*, Dr. D. Steinkopff Verlag, Darmstadt.
Cotton, F. A.: 1963, *Chemical Applications of Group Theory*, Interscience Publishers, New York.
Edmonds, A. R.: 1960, *Angular Momentum and Quantum Mechanics*, Princeton University Press, Princeton, N.J.
Fano, U. and Racah, G.: 1959, *Irreducible Tensorial Sets*, Academic Press Inc., New York.
Lepard, D. W.: 1970, *Can. J. Phys.* **48**, 1664.
Mortensen, O. S. and Koningstein, J. A.: 1968, *J. Chem. Phys.* **48**, 3971.
Placzek, G.: 1934, *Marx Handbuch der Radiologie* (2d ed.), Vol. VI, Academische Verlagsgesellschaft, Leipzig, p. 209.
Tinkman, M.: 1964, *Group Theory and Quantum Mechanics*, McGraw-Hill, New York.

CHAPTER IV

THE NORMAL RAMAN EFFECT

From the derivation of an element of the scattering tensor it is seen that the normal Raman effect is a second order effect. If the scattering particle interacts with one photon of energy $h\nu_0$, then the elements of the tensor for Rayleigh scattering are obtained from expressions which involve the wave functions of the unperturbed atom, ion or molecule (first order effect). The appropriate expressions for normal Raman scattering involve perturbed and unperturbed wave functions (second order effect), and for higher order Raman scattering (the hyper Raman effect) only perturbed wave functions play a role. The wave functions referred to in the above are of electronic origin even if the Raman transition takes place between either rotational or vibrational levels of the electronic ground state. Only for electronic Raman scattering does the (average) value of an element of $(\alpha_{\varrho\sigma})_{kn}$ occur between two electronic states (k and n) and the vibrational and rotational Raman effects may be looked upon as being electronic Raman effects because they are related to the electronic polarizability $(\alpha_{\varrho\sigma})_{kk}$. Throughout the development of Raman spectroscopy, the vibrational effect has been most important, followed by the rotational Raman effect. A considerable amount of attention was paid to these directions during the pre-laser period, but only two electronic Raman effects were reported. After the invention of the laser and its introduction as source for excitation of Raman spectra, attention shifted first to vibrational Raman scattering. Developments in rotational Raman spectroscopy were somewhat disappointing, but there has been an increase in interest in electronic Raman scattering (ERS). It is seen from the basic expressions for Raman scattering of ground states that the energy difference of the states – which are involved in the scattering process – should be smaller than the energy of the incident exciting radiation. Thus, scattering to states which are more than 5000 cm^{-1} above the initial state are not observed easily, and in general, Raman spectra are reported in the spectral range of $\sim 10\ cm^{-1}$–$5000\ cm^{-1}$. Detailed information on the position of electronic states with energy $< 5000\ cm^{-1}$ of numerous molecules, ions, or atoms, was not so abundantly available in the pre-laser period if compared to the position of rotational or vibrational levels. This is at least one reason why ERS is not so popular. Another reason is that with a few exceptions, only ions in solids have low-lying electronic levels, and laser sources are a must if Raman spectra are to be excited. Finally, there is another factor. The chemist has always been interested in the vibrational and, to a lesser degree, the rotational Raman effect from a structural point of view. But the physicist has always been interested in devicing new experiments. Generally speaking, it can be said that they concentrated their efforts on solids – in particular, single crystals – and on stimulated and hyper Raman effects. Those chemists and physicists who were interested in electronic spectroscopy had more powerful methods available for finding the position of low-lying electronic states.

Nevertheless, many features of electronic Raman scattering may be directly adapted to explain the vibrational and rotational Raman effect, and it seems worthwhile to first explain the former.

IV-1. Theory of the Electronic Raman Effect

In order to explain the various Raman effects it is necessary to investigate the expression for an element of the scattering tensor in detail. Of particular relevance is the type of wave function which should be employed in the matrix elements of the electronic dipole operator.

Electronic Raman scattering has been observed for ions in solids (and also for the NO molecule) so that in the zero order Born-Oppenheim erapproximation, the wave function of the scattering entity can be written as:

$$\Psi_{\text{tot}} = \psi_k^0(\mathbf{r})\,\varphi_{v_m}(Q_m)\,\vartheta(J, M) \qquad \text{(IV, 1-1)}$$

where $\psi_k^0(\mathbf{r})$ is the electronic wave function for the state k, $\varphi_{v_m}(\mathbf{Q}_m)$ is a vibrational wave function (v_m is a quantum number) and $\vartheta(J, M)$ is a rotational wave function. For ions in solids, the rotational part can be excluded. The wave functions $\varphi_v(\mathbf{Q})$ and the normal coordinates $\mathbf{Q}_m$ for molecules are discussed in detail in Section IV-3. Substitution of the electronic and vibrational wave functions in the expression for α results in:

$$(\alpha_{\varrho\sigma})_{k,v_m;\,n,v'_m} =$$

$$= \frac{1}{h}\sum_{r,v''_m}\frac{\langle\psi_k^0(\mathbf{r})\varphi_{v_m}(\mathbf{Q}_m)|\,e\sigma\,|\psi_r^0(\mathbf{r})\,\varphi_{v''_m}(\mathbf{Q}_m)\rangle\,\langle\psi_r^0(\mathbf{r})\,\varphi_{v''_m}(\mathbf{Q}_m)|\,e\varrho\,|\psi_n^0(\mathbf{r})\,\varphi_{v'_m}(\mathbf{Q}_m)\rangle}{h\nu_{r,\,v''_m;\,k,\,v_m} - h\nu_0} +$$

$$+ \frac{[\varrho\leftrightarrow\sigma]}{h\nu_{r,\,v'';\,nv'} + h\nu_0} \qquad \text{(IV, 1-2)}$$

The vibrational wave function $\varphi_{v_m}(\mathbf{Q}_m)$ is that of one particular vibration ($\mathbf{Q}$) of the crystal which involves movement of the scattering ion, and v_m, v'_m and v''_m are quantum numbers of levels of the vibration ($\mathbf{Q}$) associated with the levels k, n and r. The Raman transition originates in the v_mth vibrational level of the kth electronic state and terminates on the v'_mth vibrational level of the nth electronic state. The summation goes over all electronic states r and all vibrational states v''_m. The electric dipole operator $e\mathbf{r}$ can, in principle, be separated in an electronic and nuclear part. Thus, in general,

$$\sum_i e_i\mathbf{r}_i = \sum_i e_i^0\mathbf{r}_i^0 + \sum_i e_a\mathbf{Q}_a\,. \qquad \text{(IV, 1-3a)}$$

$\mathbf{Q}_a$ is the position vector of the ath nucleus, $\mathbf{r}_i^0$ the position vector of the ith electron, e_a and e_i^0 are charges of nuclei and electrons. In all applications, except when dealing with (infrared) vibrational transitions, the dipole operator is approximated with the electronic part $\sum_i e_i^0 \mathbf{r}_i^0$ so that

$$\sum_i e_i \mathbf{r}_i \approx \sum_i e_i^0 \mathbf{r}_i^0 . \tag{IV, 1-3b}$$

The denominator of the first member Equation (IV, 1-2) can then also be written as:

$$\sum_{v''} \langle \varphi_{v_m}(\mathbf{Q}_m) | \varphi_{v''_m}(\mathbf{Q}_m) \rangle \langle \varphi_{v''_m}(\mathbf{Q}_m) \varphi_{v'_m}(\mathbf{Q}_m) \rangle \times \\ \times \sum_r \langle \psi_k^0(\mathbf{r}) | e\varrho | \psi_r^0(\mathbf{r}) \rangle \langle \psi_r^0(\mathbf{r}) | e\sigma | \psi_n^0(\mathbf{r}) \rangle . \tag{IV, 1-4}$$

The numerator of the first member of (IV, 1-2) is

$$\sum_{rv''_m} \frac{1}{\nu_{r,\, v''_m;\, k,\, v_m} - \nu_0}$$

and if the spacing of the vibrational levels – of the particular vibrational mode Q – of the electronic states k, r and n are exactly identical, that is, their potential surfaces are equal, then the integral $\langle \varphi_{v_m}(\mathbf{Q}) \,|\, \varphi_{v''_m}(\mathbf{Q}_m) \rangle$ is only non-vanishing if $v_m = v''_m$. The result is that the numerator does not vary. In general, the potential surfaces are not identical, and the numerator may show a dependence on v''_m. Assuming that this variation is small, so that $v_{r,\, v''_m;\, k,\, v_m}$ can be replaced by $\bar{\nu}_{r,\, k}$, the closure theorem may be used in the space of the vibrational wave functions.* The result is that:

$$\sum_{v''} \langle \varphi_{v_m}(\mathbf{Q}_m) \,|\, \varphi_{v''_m}(\mathbf{Q}_m) \rangle \langle \varphi_{v''_m}(\mathbf{Q}_m) \,|\, \varphi_{v'_m}(\mathbf{Q}_m) \rangle = \\ = \langle \varphi_{v_m}(\mathbf{Q}_m) \,|\, \varphi_{v'_m}(\mathbf{Q}_m) \rangle . \tag{IV, 1-5}$$

The latter integral represents the Franck-Condon overlap of the vibrational wave functions of the electronic states k and n. For the element of the scattering tensor $(\alpha_{\varrho\sigma})_{k,\, v_m;\, nv'_m}$ the following expression can be written:

$$(\alpha_{\varrho\sigma})_{k,\, v_m;\, nv'_m} = \langle \varphi_{v'_m}(\mathbf{Q}_m) \,|\, \varphi_{v_m}(\mathbf{Q}_m) \rangle (\alpha_{\varrho\sigma})_{k,\, n} \tag{IV, 1-6a}$$

where

$$(\alpha_{\varrho\sigma})_{kn} = \frac{1}{h} \sum_r \frac{\langle k | e\sigma | r \rangle \langle r | e\varrho | n \rangle}{\bar{\nu}_{rk} - \nu_0} + \frac{[\varrho \leftrightarrow \sigma]}{\bar{\nu}_{rn} + \nu_0} . \tag{IV, 1-6b}$$

It is worthwhile to re-emphasize that the intensity of Raman transitions is proportional to the fourth power of the frequency of the scattered light wave. For large shifts, the intensity decreases rapidly (assuming all other factors are approximately constant). Furthermore, the electronic Raman effects presently of interest are restricted to levels with energy less than 3000 cm^{-1} above the ground state, and the Franck-Condon overlap may thus be close to unity. At the same time, transitions of the type $n,\, v'_m \neq v_m \leftrightarrow k,\, v_m$ are directly related to the overlap and consequently the intensity should be small. The conclusion may then be drawn that the electronic Raman effect with small shifts does not depend on the type of vibration of the ion in the

* Van Vleck, J. H.: 1929, *Proc. Natl. Acad. Sci.* **15**, 754.

crystal, and is related to zero order Born-Oppenheimer electronic wave functions.

The assumption that the motions of the electrons can be completely separated from nuclear motions may not be true. Vibronic coupling is a well-known effect in spectroscopy of complexes, and it seems more appropriate to define an electronic wave function $\psi_k(\mathbf{r}, \mathbf{Q})$ which is dependent on the vibrational mode $(\mathbf{Q})$. The total wave function may then be written as:

$$\Psi_{\text{tot}} = \psi_k(\mathbf{r}, \mathbf{Q}_m)\,\varphi_{v_m}(\mathbf{Q}_m). \qquad \text{(IV, 1-7)}$$

Here again the Born-Oppenheimer approximation generally is not valid if the electronic state k is degenerate. However, if the coupling of electronic and vibrational parts is small and the electronic states are non-degenerate, the Born-Oppenheimer approximation is valid. For the electronic part, the Schrödinger equation is:

$$\mathscr{H}(\mathbf{r}, \mathbf{Q}_m)\cdot\psi(\mathbf{r}, \mathbf{Q}_m) = \left[-\frac{\partial^2}{\partial r^2} + V(\mathbf{r}, \mathbf{Q})\right]\psi(\mathbf{r}, \mathbf{Q}_m) = W\cdot\psi(\mathbf{r}, \mathbf{Q}_m). \qquad \text{(IV, 1-8)}$$

The electronic Hamiltonian $\mathscr{H}(\mathbf{r}, \mathbf{Q}_m)$ also depends on the nuclear motions. If the dependence is small, then $\mathscr{H}(\mathbf{r}, \mathbf{Q}_m)$ may always be represented by a well-behaved function of $\mathbf{Q}$, and such a function can readily be expanded in a Taylor series around the point $\mathbf{Q}=0$. This is the zero point for the vibrational mode and is fixed as the equilibrium position of the nuclei. Thus:

$$\mathscr{H}(\mathbf{r}, \mathbf{Q}_m) = \mathscr{H}(\mathbf{r}, \mathbf{Q}_m = 0) + \left(\frac{\partial \mathscr{H}}{\partial Q_m}\right)_{Q_m=0} \cdot \mathbf{Q}_m + \text{higher order terms.} \qquad \text{(IV, 1-9a)}$$

Using first order perturbation theory, the expression for $\psi_k(\mathbf{r}, \mathbf{Q})$ is:

$$\psi_k(\mathbf{r}, \mathbf{Q}) = \psi_k^0 + \sum_{t \neq k} \frac{\langle \psi_t^0 | (\partial \mathscr{H}/\partial Q_m)_{Q_m=0}\, \mathbf{Q}_m | \psi_k^0 \rangle}{E_k^0 - E_t^0}\, \psi_t^0 \qquad \text{(IV, 1-9b)}$$

or

$$\psi_k(\mathbf{r}, \mathbf{Q}) = \psi_k^0 + \sum_{t \neq k} h_{tk}^{Q_m} \cdot \mathbf{Q}_m \cdot \psi_t^0 \quad \text{where} \quad h_{tk}^{Q_m} = \frac{\langle \psi_t^0 | (\partial \mathscr{H}/\partial Q_m)_{Q_m=0} | \psi_k^0 \rangle}{E_k^0 - E_t^0}$$

the superscript 0 denotes values at the point $\mathbf{Q}=0$, e.g. $\psi_k^0 = \psi(\mathbf{r}, \mathbf{Q}=0)$. With the introduction of the Born-Oppenheimer approximation, each state can be characterized by two quantum numbers, specifying the electronic and the vibrational quantum numbers respectively. By means of the expansions introduced, the matrix element $\langle \psi_{k,v_m} | e\sigma | \psi_{n,v''_m} \rangle$ can be written as:

$$\begin{aligned}\langle \psi_{k,v_m} | e\sigma | \psi_{r,v''_m} \rangle = {} & \langle \psi_k^0 | e\sigma | \psi_r^0 \rangle \langle \varphi_{v_m}(\mathbf{Q}_m) \,|\, \varphi_{v''_m}(\mathbf{Q}_m) \rangle + \\ & + \sum_{t \neq r} h_{tr}^{Q_m} \langle \psi_k^0 | e\sigma | \psi_t^0 \rangle \langle \varphi_{v_m}(\mathbf{Q}_m) \,|\, \mathbf{Q}_m \,|\, \varphi_{v''_m}(\mathbf{Q}_m) + \\ & + \sum_{t \neq k} h_{tk}^{Q_m} \langle \psi_r^0 | e\sigma | \psi_t^0 \rangle \langle \varphi_{v''_m}(\mathbf{Q}_m) | \mathbf{Q}_m | \varphi_{v_m}(\mathbf{Q}_m) \rangle + \\ & + \text{terms in } (h^Q)^2 \end{aligned} \qquad \text{(IV, 1-10)}$$

and

$$\begin{aligned}\langle\psi_{r,v''_m}|\, e\varrho\, |\psi_{n,v'_m}\rangle = {} & \langle\psi^0_r|\, e\varrho\, |\psi^0_n\rangle\, \langle\varphi_{v''_m}(\mathbf{Q}_m)\,|\,\varphi_{v'_m}(\mathbf{Q}_m)\rangle + \\ & + \sum_{t\neq n} h^{Q_m}_{tn}\, \langle\psi^0_r|\, e\varrho\, |\psi^0_t\rangle\, \langle\varphi_{v''_m}(\mathbf{Q}_m)|\, \mathbf{Q}_m\, |\varphi_{v'_m}(\mathbf{Q}_m)\rangle + \\ & + \sum_{t\neq r} h^{Q_m}_{tr}\, \langle\psi^0_n|\, e\varrho\, |\psi^0_t\rangle\, \langle\varphi_{v'_m}(\mathbf{Q}_m)|\, \mathbf{Q}_m\, |\varphi_{v''_m}(Q_m)\rangle \\ & + \text{terms in } (h^Q)^2. \end{aligned} \qquad \text{(IV, 1-11)}$$

Substitution of Equations (IV, 1-10) and (IV, 1-11) into (IV, 1-2) leads to the following expression for an element of the scattering tensor:

$$\begin{aligned}(\alpha_{\varrho\sigma})_{k,v_m;\,n,v'_m} = {} & \frac{1}{h}\sum_{rv''_m} \frac{1}{\nu_{r,v''_m;\,k,v_m} - \nu_0}\, [(M^0_\sigma)_{kr}(M^0_\varrho)_{rn}\, \langle\varphi_{v_m}\,|\,\varphi_{v_m''}\rangle\, \langle\varphi_{v''_m}\,|\varphi_{v'_m}\rangle + \\ & + \sum_{t\neq r} h^{Q_m}_{tr}\, \{(M^0_\sigma)_{kr}(M^0_\varrho)_{nt}\, \langle\varphi_{v''_m}\,|\,\varphi_{v_m}\rangle\, \langle\varphi_{v'_m}|\, \mathbf{Q}_m\, |\varphi_{v''_m}\rangle + \\ & + (M^0_\varrho)_{rn}(M^0_\sigma)_{kt}\, \langle\varphi_{v'_m}\,|\,\varphi_{v''_m}\rangle\, \langle\varphi_{v''_m}|\, \mathbf{Q}_m\, |\varphi_{v_m}\rangle\} + \\ & + \sum_{t\neq n} h^{Q_m}_{tm}\, (M^0_\sigma)_{kr}(M^0_\varrho)_{rt}\, \langle\varphi_{v_m}\,|\,\varphi_{v''_m}\rangle\, \langle\varphi_{v''_m}|\, \mathbf{Q}_m\, |\varphi_{v'_m}\rangle + \\ & + \sum_{t\neq k} h^{Q_m}_{tk}\, (M^0_\varrho)_{rn}(M^0_\sigma)_{rt}\, \langle\varphi_{v''_m}\,|\,\varphi_{v'_m}\rangle\, \langle\varphi_{v''_m}\,|\, \mathbf{Q}_m\,|\, \varphi_{v_m}\rangle + \\ & + \text{terms in } (h^Q)^2] + \\ & + \frac{\begin{bmatrix}\varrho\leftrightarrow\sigma \\ v_m \leftrightarrow v'_m\end{bmatrix}}{\nu_{r\;,v''_m;\,n,v'_m} + \nu_0}. \end{aligned} \qquad \text{(IV, 1-12)}$$

In this expression the simplification has been introduced that all wave functions are real so that $(M^0_\varrho)_{rk} = (M^0_\varrho)_{kr}$ and also $h^{Q_m}_{tr} = h^{*Q_m}_{tr}$. Also included is the assumption that the electronic states k and n show vibronic coupling but not the intermediate state r. If the latter condition is not satisfied then additional contributions arise from the constants $h^{Q_m}_{tn}$ and $h^{Q_m}_{tk}$. Finally a summation has been introduced over all $\mathbf{Q}_m$. If the closure rule is applied in the space of the vibrational wave functions then the resulting expression is:

$$\begin{aligned}(\alpha_{\varrho\sigma})_{k,v_m;\,n,v'_m} = {} & \frac{1}{h}\sum_r \left[\frac{(M^0_\sigma)_{kr}(M^0_\varrho)_{rn}}{\bar\nu_{rk} - \nu_0} + \frac{[\varrho\leftrightarrow\sigma]}{\bar\nu_{rn} + \nu_0}\right] \langle\varphi_{v_m}(\mathbf{Q}_m)|\, \varphi_{v'_m}\, |\mathbf{Q}_m\rangle + \\ & + \langle\varphi_{v_m}(\mathbf{Q}_m)|\, \mathbf{Q}_m\, |\varphi_{v'_m}(\mathbf{Q}_m)\rangle \left[\sum_{t,Q} \frac{h^{Q_m}_{tr}((M^0_\sigma)_{kr}(M^0_\varrho)_{nt} + (M^0_\varrho)_{rn}(M^0_\sigma)_{kt})}{\bar\nu_{r,k} - \nu_0}\right. \\ & \left. + \frac{[\varrho\leftrightarrow\sigma]}{\bar\nu_{rn} + \nu_0}\right] \\ & + \text{terms in } h^Q_{tn}, h^Q_{tk} \ldots. \end{aligned} \qquad \text{(IV, 1-13)}$$

The zero order term is, of course, identical to Equation (IV, 1-6). The second order term contains several interesting features. Again assuming that the potential surfaces for the vibrational mode $\mathbf{Q}_m$ of the electronic states k and n are similar, the matrix element $\langle \varphi_{v_m} | \mathbf{Q}_m | \varphi_{v'_m} \rangle$ is non-vanishing if $v'_m = v_m \pm 1$. This follows directly from the vibrational wave functions as will be shown later on (see section on vibrational Raman effect). The first order terms are apparently related to the transitions n, $v_m \pm 1 \leftrightarrow k$, v_m, but the contributions of these transitions to the electronic Raman spectrum are necessarily small because the terms $h_{tr}^{Q_m}$ etc. are small. If the potential surfaces of the vibrational mode $\mathbf{Q}_m$ are not equal for the states k and n, then deviation may occur. Under appropriate conditions, the transitions n, $v_m \pm 2 \leftrightarrow k$, v_m may appear in the Raman spectrum. However, by and large, the strongest lines of the electronic Raman spectrum are the pure electronic transitions n, $v'_m = v_m \leftrightarrow k$, v_m. From the expression for an element of the scattering tensor (Equation (IV-13)), it follows that $(\alpha_{\varrho\sigma})_{kn} \neq$ $\neq (\alpha_{\sigma\varrho})_{kn}$. The conclusion is reached that the tensor can become partly antisymmetric. There is no way to predict the occurrence of a complete antisymmetric tensor i.e. $\alpha_{\varrho\sigma} = -\alpha_{\sigma\varrho}$.

IV-2. Electronic Levels and Selection Rules

Trivalent lanthanide ions have the electronic configuration $[\mathrm{Xe}](4f)^n$, where $n = 1$–14, and the configuration for the transition metal ions of interest here is $[\mathrm{Ar}]\,(3d)^n$, where $n = 1$–10. These electronic configurations generate a larger number of electronic levels, and for rare earth ions, they are found in the energy range of $0\ \mathrm{cm}^{-1}$–$60000\ \mathrm{cm}^{-1}$. The Russell-Saunders coupling scheme may sometimes be employed to find the term schemes of the lanthanides. Thus – (i) the individual orbital quantum numbers l_i of the electrons are combined vectorially to yield $\mathbf{L}$, and (ii) all individual spins s_i are added to give a total spin quantum number $\mathbf{S} = \sum_i s_i$. The total angular momentum $\mathbf{J}$ is obtained by adding the vectors $\mathbf{L}$ and $\mathbf{S}$ so that:

$$J = \mathbf{L} + \mathbf{S},\ \mathbf{L} + \mathbf{S} - 1, \ldots |\mathbf{L} - \mathbf{S}|\,. \tag{IV, 2-1}$$

Electronic configurations give rise to spectroscopic terms and they are labelled with the quantum numbers S, L and J. The notation for the terms is ${}^{2S+1}L_J$ and S, P, D, F... terms are related to $\mathbf{L} = 0$, 1, 2, 3... respectively. The spin multiplicity is given by $2\mathrm{S}+1$. The energy difference of the terms can in principle be calculated if the Schrödinger equation for the atoms or ions is solved. The Hamiltonian of a free atom or ion is:

$$\mathscr{H} = \nabla^2 - \sum_j \frac{Ze^2}{r_j} + \sum_{i \neq j} \frac{e_i e_j}{{}_j r_i} + \lambda(\mathbf{L}\cdot\mathbf{S})\,. \tag{IV, 2-2}$$

∇^2 is the Laplacian and represents the kinetic energy and Ze^2/r_j is the attractive force between the jth electron and the nucleus with a charge of $+Ze$. The repulsive forces between the electrons are represented by the third member. The fourth member is the spin-orbit interaction.

As a result of the orbital motion of the electron, a magnetic field is generated at the place of the nucleus, and the magnetic moment of the spin interacts with this magnetic field. Thus the spin-orbit interaction is of magnetic origin. The Schrödinger equation can be solved for a one-electron system but some problems do arise if two or more electrons are involved. It is found that the one electron hydrogen type wave functions may be used to construct the wave function of the atom (or ion), but these wave functions can be admixtures of the one electron functions. The wave function of a $^4I_{9/2}$ state of the Nd^{3+} ion is actually a composition of wave function of $^4F_{9/2}$, $^4G_{9/2}$, $^2H_{9/2}$ states. The quantum number J is still a good quantum number, but the admixture shows that this is definitely not so for L.

The trivalent Europium ion has six $4f$ electrons and the positions of the lowest J manifolds are given in Figure IV-1. There are, however, many other terms at higher energies. The energy difference of the J manifolds inside a term is a measure of the

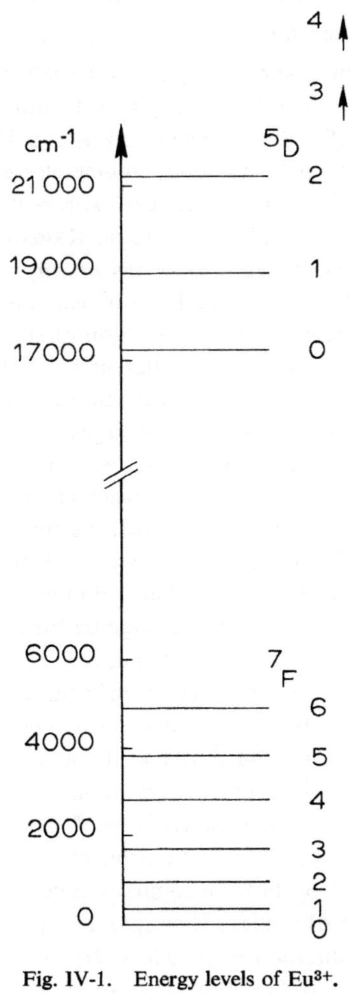

Fig. IV-1. Energy levels of Eu^{3+}.

spin-orbit coupling and the difference of ^5D and ^7F is related to the repulsive forces between the f electrons. Theory predicts that differences of ^7F_J manifold levels are proportional to the value of J of the higher level (the Landré interval rule). However, even without an a priori knowledge of the appropriate labelling of the energy states it is possible to make already some remarks about selection rules for electronic Raman transitions in considering the parity of the electronic states. The parity of electronic states is exclusively determined by the value of the orbital quantum number of the electron. All electronic states originating from the interaction of spin and orbital movements of an 1s-electron are of *even* parity because the wave function of an s orbital ($l=0$) is continuously positive. On the other hand, the wave function of a p-orbital ($l=1$) changes sign upon reflection through the origin, and, per definition, all electronic states originating from the configuration $(p)^1$ or $(p)^n$ are of *odd* parity. Consequently, the wave functions of states of the lanthanide ions are of even parity. Also, the Raman operator $\hat{\alpha}_{\varrho\sigma}$ is an even operator because $\varrho \xrightarrow{\hat{i}} -\varrho$ and $\sigma \xrightarrow{\hat{i}} -\sigma$ thus $\varrho\sigma \xrightarrow{\hat{i}} \sigma\varrho$. Hence, the integral $\langle\psi_k|\,\hat{\alpha}_{\varrho\sigma}\,|\psi_n\rangle \neq 0$ if ψ_k^* and ψ_n have the same parity. If one of the wave functions is of even parity and the other odd, then the value of the product $\psi_k\hat{\alpha}_{\varrho\sigma}\psi_n$ for a small volume element ΔV is equal but opposite in sign to the value of this product for a small but equal volume element ΔV^1 related to ΔV by a reflection through the origin. This process can be repeated for all space so that the final result is $\langle\psi_k|\,\hat{\alpha}_{\varrho\sigma}\,|\psi_n\rangle=0$. Thus, electronic Raman transitions between levels arising from $(4f)^n$ and also $(3d)^n$ configurations are formally allowed! Unfortunately, Raman scattering of free transition metal or rare earth ions has not yet been found, but information is available concerning transition of such ions in crystals.

The Hamiltonian for ions in solids is different from the Hamiltonian of the free ion. The electrons of the centrally located transition or lanthanide ion will experience an electrostatic potential as a result of the charges of the nearest neighbours. Such neighbours are sometimes taken as point charges and a calculation is made of the potential which such charges produce at the place of the d or f electrons of the ions. The crystal field potential H_{CF} produces a small perturbation for lanthanide ions in solids. That is, the spin-orbit interaction term in the Hamiltonian is larger than H_{CF} and J remains a good quantum number. This situation is reversed for the transition metal ions and the H_{CF} is looked upon as a large perturbation of the free ion. In that case, even J is no longer a good quantum number. The aspect of crystal field and ligand field theory which is of importance to this chapter is not the exact calculation of the position of electronic states but rather the symmetry of the crystal field at the place of the ions. The surrounding ions may be arranged in such a way that the element of highest symmetry is a four-fold rotation axis in the z-direction. It is now convenient to refer the symmetry considerations of electronic states etc. to this axis. Using the language of Quantum Mechanics, this axis is considered to be *the* axis of quantization. The projection of the total angular momentum J on this axis can assume only certain fixed (quantized) values. There are $2J+1$ of such projections and they are labelled with the quantum number J_z. The energy levels of a *free* ion labelled with the quantum number L, S and J are thus $2J+1$-fold degenerate, and this degeneracy can be removed by the crystal field. It should be obvious that the higher the symmetry of the environment, the more the situation approaches that of the free ion; that is, the degeneracy remains intact. Examples for different J manifolds are given in Table IV-1. For crystal fields of low symmetry, the degeneracies can be completely removed.

The crystal field perturbations for transition metal ion are quite different from those of the lanthanides. In the so-called weak field case, $\lambda(\mathbf{L}\cdot\mathbf{S})<H_{CF}<e^2/r_{ij}$. The crystal field H_{CF} is small with respect to the repulsive forces of the electrons, but larger than the spin-orbit coupling. J is no longer a good quantum number, but L and S retain their validity. In the strong field case, $H_{CF}<e^2_{r_{ij}}$, and even L is not a good quantum number. It is only possible here to remember that d electrons are making up the electronic configurations. The spacial orientation of the five d orbitals is described by the appropriate spherical harmonic functions $Y^2_m(m=-2, -1, 0, 1, 2)$ – (see Table III-7). For the free ion, all five orbitals have the same energy. The symmetry (and degeneracy) of the orbitals in surroundings of different symmetry can thus be determined using the transformation properties of the spherical harmonics as outlined in Chapter III.

A similar procedure can be followed in the weak field case. The configuration $(3d)^2$ gives to ^3P and ^2F states, and the splitting of these states is obtained by the transformation properties of the spherical harmonics Y^1_m and Y^3_m. The results given in Table IV-1 are obtained in a similar way by replacing l with J and m with J_z in the spherical harmonics Y^l_m. The selection rules for electronic Raman transitions of lanthanide ions can now be refined by making use of the irreducible tensor components α^K_Q and as an example may serve the case of Eu^{3+}. For the transition $F_1\leftarrow F_0$ (see Figure IV-1), the ground state ($J=0$) transforms according to the totally symmetric species of the point group which describes the symmetry at the place of the ions. The three levels of the $^7F_1$ manifold transform like the spherical harmonics Y^l_m $(m= =-1, 0, +1)$. For a Raman process to become allowed, it is thus a prerequisite that the components α^K_Q transform like the Y^l_m's, and these components are $\alpha^1_Q(Q=-1, 0, +1)$. The antisymmetric part of the Cartesian tensor is contributing to the Raman intensity! In the same way, it can be shown that the components for the $^7F_2\leftarrow{}^7F_0$ transition are $\alpha^2_Q(Q=-2, -1, 0, +1, +2)$ and $^7F_3\leftarrow{}^7F_0$ is not allowed. The selection. rule is thus $\Delta J\leqslant 2$. Replacing the quantum number J by L and J_z by m yields the selection rule $\Delta L\leqslant 2$ and also $\Delta S=0$. The $^7F_1$–$^7F_0$ transitions for Eu^{3+} in $Eu_3Ga_2(GaO_4)_3$ have been studied by Koningstein *et al.** and in agreement with theory complete antisymmetric tensors were measured. The example of the Europium ion is a particularly nice example of the occurrence of an antisymmetric scattering tensor. However, this is not the only interesting feature of the electronic Raman effect. For ions with an odd number of electrons, all crystal field levels are two-fold degenerate even if the symmetry of the environment is monoclinic. This degeneracy can only be removed if the sample is exposed to a magnetic field.

The value of the quantum number J of such ions or atoms is always a half-integer, and this gives rise to some problems in finding the symmetry properties of the states.

The behaviour of wave functions under the operations of a group is determined

* Dabrowski, J., Grünberg, P., and Koningstein, J. A.: 1972, *J. Chem. Phys.* **56**, 1264.

TABLE IV-1

Splitting of levels with a given $l(L)$ or $j(J)$ in surroundings of different symmetry

	Symmetry of surroundings and iniducible representations		
j or l	O_h	D_{4h}	D_2
0	A_{1g}	A_{1g}	A
1	T_{1u}	$A_{2u} + E_u$	$B_1 + B_2 + B_3$
2	$E_g + T_{2g}$	$A_{1g} + B_{1g} + B_{2g} + E_g$	$2A + B_1 + B_2 + B_3$
3	$A_{2u} + T_{1u} + T_{2u}$	$A_{2u} + B_{1u} + B_{2u} + 2E_u$	$A + 2B_1 + 2B_2 + 2B_3$
4	$A_{1g} + E_g + T_{1g} + T_{2g}$	$2A_{1g} + A_{2g} + B_{1g} + B_{2g} + 2E_g$	$3A + 2B_1 + 2B_2 + 2B_3$
5	$E_u + 2T_{1u} + T_{2u}$	$A_{1u} + 2A_{2u} + B_{1u} + B_{2u} + 3E_u$	$2A + 3B_1 + 3B_2 + 3B_3$
6	$A_{1g} + A_{2g} + E_g + T_{1g} + 2T_{2g}$	$2A_{1g} + A_{2g} + 2B_{1g} + 2B_{2g} + 3E_g$	$4A + 3B_1 + 3B_2 + 3B_3$

For states determined by the total angular moment J or the total orbital quantum number L, the subscripts g and u are exclusively given by the value of l of the single electron wavefunction.

solely by function $\phi(\varphi)$ – see Equation (III, 5-10). Both the radial function $R_{nl}(r)$ and the function $\theta_m^l(\theta)$ are invariant under the operations. The solution of the Schrödinger equation yields the result:

$$\phi_m(\varphi) = N_\varphi \, e^{im\varphi} \tag{IV, 2-3a}$$

where $N(\varphi)$ is a normalization constant. For a rotation of ϕ degrees around the z-axis which is the axis of quantization, the angle φ will be carried over in $\varphi + \phi$, and the function $\varphi_m(\varphi)$ becomes:

$$\phi_m(\varphi) = N_\varphi \, e^{im(\varphi + \phi)}. \tag{IV, 2-3b}$$

The value of the quantum number m goes from $+l$, $l-1, \ldots -l$ and under an n-fold rotation the wave function $\phi_m(\varphi)$ can be carried over into itself or another of the set of $2l+1$ functions. The rotation is thus given by the matrix

$$\begin{pmatrix} e^{il\varphi} & & \\ & e^{i(l-1)\varphi} & \\ & & \ddots \\ & & & e^{-il\varphi} \end{pmatrix}$$

and the character χ of this rotation (ϕ) is thus:

$$\begin{aligned} \chi(\phi) &= e^{il\phi} + e^{i(l-1)\phi} + \cdots + e^{-il\phi} \\ &= e^{-il\phi}\left(e^{2il\phi} + e^{i\phi(2il-1)} + \cdots + 1\right) \\ &= e^{-il\phi}\frac{\left(1 - e^{i\phi(2l+1)}\right)}{1 - e^{i\varphi}} \\ &= \frac{e^{1/2i\phi}\left[e^{-(l+1/2)i\phi} - e^{(l+1/2)i\phi}\right]}{1 - e^{i\varphi}} \\ &= \frac{e^{-(l+1/2)i\phi} - e^{(l+1/2)i\phi}}{e^{-1/2i\varphi} - e^{1/2i\varphi}} \\ &= \frac{\sin\left[\left(l + \frac{1}{2}\right)\phi\right]}{\sin\frac{1}{2}\phi} \end{aligned} \tag{IV, 2-4}$$

The characters for some rotations are thus:
$\chi(\hat{C}_2) = (-1)^l$ and $\chi(\hat{C}_4) = +1$ for $l=0, 1, 4, 5$ and $\chi(\hat{C}_4) = -1$ for $l=2, 3, 6, 7 \ldots$.

The quantum numbers m and l are the magnetic and orbital quantum numbers of hydrogen type wave functions, but in the situation of more electron atoms, the important quantum numbers are J_z and J respectively. The character of rotation through an angle ϕ is then represented by:

$$\chi(\phi) = \frac{\sin\left(J + \frac{1}{2}\right)\phi}{\sin\frac{1}{2}\phi} \tag{IV, 2-5}$$

for J integer $\chi(\phi) = \chi(\phi + 2\pi)$, but for J half-integer $\chi(\phi) = -\chi(\phi + 2\pi)$.

The occurrence of half-integer values of the quantum number J is solely due to the fact that the spin of the electrons is taken into account. Such a situation cannot occur in the case treated above for the quantum numbers l or m only.

From the character $\chi(\phi)$ for integer values of J it follows that the system goes over into itself if the rotation amounts to 360°, but this is not so for half-integer values of J. Apparently all characters change sign upon rotations of 360. On the other hand, it follows that a rotation of 4π is equivalent to the identity operation $\hat{E}$.

The fiction is now introduced that the molecule does not go into itself upon a rotation of 2π around the z-axis but only after a rotation of 4π or $2 \times 360°$. A new group element, the rotation by 2π is introduced and this element, in combination with the normal symmetry elements, is used to construct new groups – the so-called double groups. Character tables and transformation properties of wave function, electric dipole operators and the Raman tensor are obtained in the normal way.

Presently, Raman transition of electronic levels which belong to representation of double groups have been found for rare earth and transition metal ions. As an example, the trivalent ytterbium ion may be chosen. If this ion is experiencing a crystal field of D_2 symmetry, then the $^2F_{7/2}$ and $^2F_{5/2}$ manifolds of the [Xe] $(4f)^{13}$ configuration are split into four and three double degenerate crystal field levels. In order to find the representations of these levels, it is necessary to employ the double group D_2'. The character table of this group is given in Table IV-2. It is found that the degenerate electronic levels belong to the 'double' valued representation $E_{1/2}$. It is readily

TABLE IV-2

Character table of the double group D'_2.

D_2	E	R	$C_2(z)$	$C_2(y)$	$C_2(x)$	Raman activity
A	1	1	1	1	1	$\alpha_{xx}, \alpha_{yy}, \alpha_{zz}$
B_1	1	1	1	-1	-1	$\alpha_{xy} + \alpha_{yx}, \alpha_{xy} - \alpha_{yx}$
B_2	1	1	-1	1	-1	$\alpha_{zx} + \alpha_{xz}, \alpha_{zx} - \alpha_{xz}$
B_3	1	1	-1	-1	1	$\alpha_{yz} + \alpha_{zy}, \alpha_{yz} - \alpha_{zy}$
$E_{1/2}$	2	-2	0	0	0	None

The levels of a $J = \frac{7}{2}, \frac{5}{2}$ manifold transform according to the double valued representation $E_{1/2}$.

verified that the product $E_{1/2} \times E_{1/2}$ contains all the 'single' valued representations A, B_1, B_2 and B_3. These single valued representations are identical to those of the normal group D_2. From the character table for D_2' it is seen that for the electronic Raman transition, the following terms contribute to the intensity:

$$\alpha_{xx}, \alpha_{yy}, \alpha_{zz},$$
$$\alpha_{xy} + \alpha_{yx}, \alpha_{xy} - \alpha_{yx},$$
$$\alpha_{xz} + \alpha_{zx}, \alpha_{xz} - \alpha_{zx},$$
$$\alpha_{yz} + \alpha_{zy}, \alpha_{yz} - \alpha_{zy}.$$

The importance of the occurrence of these tensors becomes evident if a rare earth is placed on the site of D_2 symmetry with an even number of $4f$ electrons. The electronic levels now belong to the representations A, B_1, B_2 and B_3, and the tensors of the Raman transitions belong necessarily to the same representations. Thus, only *one* of the four possible tensors given above is allowed. Double groups are also encountered in the Raman effects of transition metal ions. For the divalent cobalt ion in a site of D_3 symmetry, the low lying electronic levels belong to the two double-valued representations of the double group D_{3d}. The Raman tensors of the transitions are thus contained in the products of the representations: $E_{1/2} \times E_{1/2} = A_1 + A_2 + E$; $E_{3/2} \times E_{3/2} = 2A_1 + 2A_2$; $E_{1/2} \times E_{3/2} = E + E$. The tensors for the transitions $E_{1/2} \leftrightarrow E_{1/2}$; $E_{3/2} \leftrightarrow E_{3/2}$ and $E_{1/2} \leftrightarrow E_{3/2}$ thus are all different. Hence, double groups are important for electronic Raman transitions, and the character tables of these groups are therefore given in Appendix I.

The spin-orbit coupling is strong for the lanthanide ions and J remains a good quantum number, even if the ions are imbedded in crystals. Such is not the case for the transition metal ions and the orbital motion of the d-electrons is 'quenched' in the strong field approximation. What is left over is the spin quantum number $S = \sum_i s_i$. The angular momentum due to the spin only is also represented by an axial vector. Such a vector does not change sign upon inversion through the origin. Thus for point groups with an inversion center, the species of the spin states are always g(gerade). For a total spin $S = 1$, there are three substates given by the projections $M_S = 0, \pm 1$ and the symmetry of these states follow from Table IV-1 by replacing l by M_S and keeping in mind that all subscripts should be g. The species of the spin states for some other important point groups are also given in the Appendix I. The symmetry of electronic states for situations between the weak and strong field approximation can always be obtained by taking the product of the species of spin and orbital wave functions. But as a result of the multiplication rule: $g \times g = g$, $g \times u = u \times g = u$, the appropriate subscript of the species is always determined by the value of the orbital quantum number (see also the discussion on the parity of states in the above).

IV-3. Normal Modes, Normal Coordinates and Vibrational Wave Functions

Of importance to the theory of the normal Raman effect are the concepts of normal modes and normal coordinates. In this section a discussion of these concepts is given, and in doing so, it shall be assumed that the total wave function Ψ_{tot} can be separated into electronic, vibrational and rotational wave functions. It was shown in section IV-1 that the separation of nuclear and electronic motions is not always straightforward, but we shall not treat the effect of vibronic interaction here. Furthermore, it should be pointed out that when a proper coordinate system is used, the vibrational part can be separated into a part which deals with the translations of the particle, and into another part with the pure vibrations only.

The vibrational motion of a non-linear molecule is, in general, quite complicated. But this motion can be represented as a superposition of several normal vibrations. These normal modes for a non-linear triatomic molecule are shown in Figure IV-2. The important questions to be answered here are related to (i) the number of such normal modes for a polyatomic molecule, and (ii) the symmetry of these modes. It is easily seen that the normal modes of Figure IV-2 possess symmetry. Here again some knowledge of group theory is most valuable. Every atom of a molecule can, in principle, move in three independent directions, and the displacements can be indicated with the Cartesian coordinates $x_1, y_1, z_1, x_2, y_2, z_2 \ldots x_n, y_n, z_n$. Thus there are a total of $3N$ degrees of freedom. However, it may happen that all atoms move simultaneously in the x-direction, y-direction or z-direction. If this is the case,

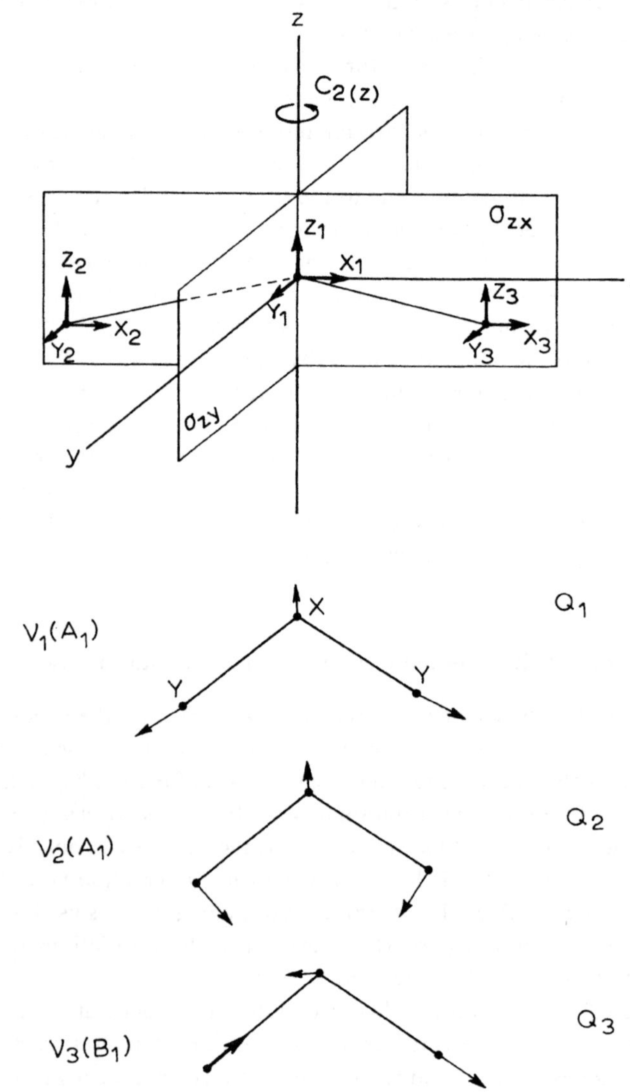

Fig. IV-2. Geometry of an XY_2 molecule of symmetry C_{2v} and the three normal modes of vibration.

then the whole molecule moves along x, y or z direction respectively. These motions are not really vibrations but rather translations. A non-linear molecule can also rotate, and there are three rotations. They are related to particular combinations of the displacement coordinates. There are thus a total of $3N$-6 (for a linear molecule $3N$-5) degrees of freedom left over for the vibrations of the molecule. A molecule XY_2 thus has $3\times 3-6=3$ vibrational modes. Another approach to the analysis of vibrations of molecules is obtained if the displacement vectors x, y, z etc. are subjected to the symmetry operations of the point group to which the molecule belongs. For example, for the XY_2 molecule (symmetry C_{2v}) there are four symmetry elements: E, C_2, $\sigma_v(zx)$ and $\sigma_v(zy)$. The two-fold rotation axis coincides with the z-axis and goes through the X-atom. It carries atom Y_1 over into Y_2, and this operation is represented by off-diagonal elements only in the transformation matrix – see Table IV-3. The character of the matrix is thus directly related to the behaviour of the displacement vectors of the X atom. Thus $\chi(\hat{C}_2)=-1$. Other characters are $\chi(E)=9$; $\chi(\sigma_v(zx))=3$ and $\chi(\sigma_v(zy))=1$. This representation is reducible, and the irreducible representations are: $\Gamma=3A_1+1A_2+3B_1+2B_2$.

In the activity representation of the point group C_{2v}, it is indicated that translations along Z, X and Y directions are contained in the species A_1, B_1 and B_2 respectively. Rotations around these axes transform* like the species A_2, B_1 and B_2, and the vibrational modes thus transform according to the representations $A_1\,(2\times)$ and $B_2\,(1\times)$. The direction of the displacement vectors – and their relative magnitudes – of the atoms can now be constructed for the various normal coordinates by considering the symmetry conditions plus the condition that all vectors should add up to a total resultant vector with a magnitude equal to zero. The latter condition implies that the molecule cannot vibrate and translate simultaneously but carries out only vibrational motions. For molecules which belong to other point groups it is sometimes observed that two or more normal modes transform according to a degenerate representation. But the above method for finding the normal modes is identical.

TABLE IV-3

Transformation matrix of displacement coordinates of a molecule XY_2 of symmetry C_{2v} for the symmetry operation $\hat{C}^2(z)$.

	x_1	y_1	z_1	x_2	y_2	z_2	x_3	y_3	z_3
x_1	−1	0	0	0	0	0	0	0	0
y_1	0	−1	0	0	0	0	0	0	0
z_1	0	0	1	0	0	0	0	0	0
x_2	0	0	0	0	0	0	−1	0	0
y_2	0	0	0	0	0	0	0	−1	0
z_2	0	0	0	0	0	0	0	0	1
x_3	0	0	0	−1	0	0	0	0	0
y_3	0	0	0	0	−1	0	0	0	0
z_3	0	0	0	0	0	1	0	0	0

$\chi(\hat{C}_2(z)) = -1.$

* See page 68.

Using group theory, it is possible to determine the normal modes, but it is rather obvious that information does not become available on the frequency of oscillation of the atoms about their equilibrium positions. Such information may be obtained by applying laws of classical mechanics to small vibrations.

Newton's equations for a conservative system are:

$$\begin{aligned}\frac{\mathrm{d}}{\mathrm{d}t}\frac{\partial T}{\partial \dot{x}_i}+\frac{\partial V}{\partial \dot{x}_i}&=0\\ \frac{\mathrm{d}}{\mathrm{d}t}\frac{\partial T}{\partial \dot{y}_i}+\frac{\partial V}{\partial \dot{y}_i}&=0\\ \frac{\mathrm{d}}{\mathrm{d}t}\frac{\partial T}{\partial \dot{z}_i}+\frac{\partial V}{\partial \dot{z}_i}&=0.\end{aligned} \qquad \text{(IV, 3-1)}$$

Here, $\dot{x}_i = \mathrm{d}x_i/\mathrm{d}t$, where x_i is the coordinate of the ith particle, T is the kinetic energy and V is the potential energy. For a molecule with N atoms, there are $3N$ of such equations. For the kinetic energy T, the following expression can be written:

$$2T = \sum_{i=1}^{N} m_i \left[\left(\frac{\mathrm{d}\Delta x_i}{\mathrm{d}t}\right)^2 + \left(\frac{\mathrm{d}\Delta y_i}{\mathrm{d}t}\right)^2 + \left(\frac{\mathrm{d}\Delta z_i}{\mathrm{d}t}\right)^2\right]. \qquad \text{(IV, 3-2)}$$

Δx_i, Δy_i and Δz_i are the deviations from the equilibrium position of the ith atom. This expression can be simplified if the mass-weighted coordinates q_i are introduced. They are related to the ordinary coordinates as follows:

$$\begin{aligned}&q_1 = \sqrt{m_1}\Delta x_1, \quad q_2 = \sqrt{m_1}\Delta y_1, \quad q_3 = \sqrt{m_1}\Delta z_1,\\ &q_4 = \sqrt{m_2}\Delta x_2 \ldots \quad q_{3N} = \sqrt{m_n}\Delta z_n.\end{aligned} \qquad \text{(IV, 3-3)}$$

The kinetic energy now assumes the form:

$$2T = \sum_{j=1}^{3N} \dot{q}_j^2. \qquad \text{(IV, 3-4)}$$

The potential energy V will also be a function of the displacements, and for small displacement V may be expressed in a power series in the coordinates q_i:

$$2V = 2V_0 + 2\sum_{j=1}^{3N}\left(\frac{\partial V}{\partial q_j}\right)_0 q_j + \sum_{j,l=1}^{3N}\left(\frac{\partial^2 V}{\partial q_j \partial q_l}\right)_0 q_j q_l + \ldots \qquad \text{(IV, 3-5a)}$$

$$= 2V_0 + 2\sum_{j=1}^{3N} f_j q_j + \sum_{j,l=1}^{3N} f_{jl} q_j q_l \qquad \text{(IV, 3-5b)}$$

where $f_j = (\partial V/\partial q_j)_0$. For vibrations of the type depicted in Figure IV-2, it is seen that the kinetic energy is at a maximum if the atoms return to the equilibrium positions. It can be shown that the potential energy is at a minimum; thus

$$\left(\frac{\partial V}{\partial q_j}\right)_0 = 0 \quad j = 1, 2, \ldots 3N.$$

Hence:

$$2V = \sum_{j,l=1}^{3N} f_{jl} q_j q_l + \ldots \tag{IV, 3-5c}$$

where

$$f_{jl} = \left(\frac{\partial^2 V}{\partial_j q\, \partial q_l}\right)_{0,0}$$

For small amplitudes of the vibrations, the power series need only be considered up to the second-order terms. Substitution of Equation (IV, 3-5c) and Equation (IV, 3-4) into Equation (IV, 3-1) results in:

$$\ddot{q}_l + \sum_{j=1}^{3N} f_{jl} q_j = 0 \quad l = 1, 2 \ldots 3N. \tag{IV, 3-6a}$$

It is seen that Equations (IV, 3-6) represent $3N$ simultaneous second-order differential equations. One possible solution is:

$$q_j = A_j \cos(\sqrt{\lambda}\, t + \varphi) \tag{IV, 3-6b}$$

where φ is a phase factor and A_j and λ are constants. Substitution of Equation (IV, 3-6b) in Equation (IV, 3-6a) results in:

$$\sum_{j=1}^{3N} (f_{jl} - \delta_{jl}\lambda)\, A_j = 0 \quad l = 1, 2 \ldots 3N. \tag{IV, 3-6c}$$

δ_{jl} is the Kronecker delta symbol and this equation represents a set of simultaneous homogenous linear algebraic equations in the $3N$ unknowns A_j. The solution for such a set has already been discussed (see chapter II, 1). There are $3N$ eigenvalues and they are indicated with $\lambda_m (m=1, 2 \ldots 3N)$. For every eigenvalue λ_m there are also $3N$ constants A_{jm}. It is not possible to find unique values for the individual A_{jm}'s, but only their ratios. Thus it is most convenient to arbitrarily equate one of the constants to unity, for instance $A_{lm} = 1$, and the set of amplitudes is then designated by A^1_{jm}. A more interesting solution is obtained by introducing the constants B_{jm}, where

$$B_{jm} = \frac{A^1_{jm}}{(\sum_j A^1_{jm})^{1/2}} \quad \text{and} \quad \sum_j B^2_{jm} = 1 \tag{IV, 3-7a}$$

so that:

$$A_{jm} = k_m B_{jm} \tag{IV, 3-7b}$$

where the constants k_m are determined by the values of q_i.

Returning to Equation (IV, 3-6b), it is seen that each atom of the molecule is oscillating about its equilibrium position with a simple harmonic motion. This is because Equation (IV, 3-6b) is similar to expressions obtained for harmonic oscillators. From such a comparison it follows that the frequency of oscillation is $\nu = (\frac{1}{2}\pi)\sqrt{\lambda}$, and the amplitude is represented by A_j. Frequencies, amplitudes and phases for the case that λ equals the eigenvalues λ_m are then designated with the values $\nu_m = (\frac{1}{2}\pi)\sqrt{\lambda_m}$, A_{jm} and φ_m respectively. It is important to note that the frequency of oscillation of all atoms of the molecule is now equal, and the oscillators are in phase. However, the amplitudes A_{jm} are generally different. The normal modes of vibration of a molecule are, in fact, defined by these properties, and the symmetry aspects were already discussed in the above.

The laws of classical mechanics do not apply, however, to atomic or subatomic systems, and the real solution pertaining to amplitudes and frequencies of oscillation of normal modes is obtained if the Schrödinger wave equation is solved. The Hamiltonian for a system of vibrating particles is related to the classical expressions for kinetic and potential energy. In general, the kinetic energy may be written as:

$$T = \tfrac{1}{2}m\dot{x}^2 = \frac{1}{2m}\,p_x^2$$

where $p_x = m\dot{x}$, but for systems where quantum mechanics is applicable, the linear momentum p_x is replaced by the differential operator $h/2\pi i\,\partial/\partial x$. The Hamiltonian $\mathscr{H} = \frac{1}{2}mp_x^2 + V(x) = W$, where W is the energy which is replaced by $-h/2\pi i\,\partial/\partial t$. The kinetic energy given by Equation (IV, 3-5) is related to the mass-weighted coordinate q_j, but the potential energy V – see Equation (IV, 3-6c) – is related to a product of two mass-weighted coordinates. Consequently, the Hamiltonian would not be related to one particular coordinate and, apparently, mass-weighted coordinates are not suitable to solve the Schrödinger wave equation for a system of vibrating particles. The coordinates which should be employed are called the normal coordinates Q and they are related to the mass-weighted coordinates q in the following way:

$$Q_m = \sum_{j=1}^{3N} a_{mj} q_j \quad m = 1, 2 \ldots 3N\,. \qquad \text{(IV, 3-8a)}$$

The a_{mj} coefficients are so chosen that the kinetic energy becomes $2T = \sum_{m=1}^{3N} \dot{Q}_m^2$ and the potential energy $2V = \sum_{m=1}^{3N} \Lambda_m Q^2$. It is seen that the potential energy does not involve cross products of normal coordinates. Of paramount importance is a knowledge of the 'a' coefficients and the values of Λ_m. There are $3N$ equations for Q_m in terms of the coefficients a_{mj} (where $j = 1, 2 \ldots 3N$), but there are also $3N$ equations for q_j in terms of the coefficients b_{jm} so that:

$$q_j = \sum_{m=1}^{3N} b_{jm} Q_m \quad j = 1, 2 \ldots 3N \qquad \text{(IV, 3-8b)}$$

and it is said that Equation (IV, 3-8b) is the inverse of Equation (IV, 3-8a). The set of linear algebraic equations which connects the two sets of coordinates q_j and Q_m is called a linear transformation. By substituting Equation (IV, 3-8b) in (IV, 3-8a) the following result is obtained:

$$\sum_{j=1}^{3N} a_{mj} b_{jn} = \delta_{mn} . \qquad \text{(IV, 3-8c)}$$

But by substituting Equation (IV, 3-8a) in (IV, 3-8b) the result is:

$$\sum_{m=1}^{3N} b_{jm} a_{ml} = \delta_{jl} . \qquad \text{(IV, 3-8d)}$$

Newton's eqation of motion in terms of the normal coordinates can be written as:

$$\frac{\mathrm{d}}{\mathrm{d}t} \frac{\partial T}{\partial Q_m} + \frac{\partial V}{\partial Q_m} = \ddot{Q}_m + \Lambda_m Q_m = 0 \quad m = 1, 2 \ldots 3N \qquad \text{(IV, 3-9a)}$$

and the solution of the second-order differential equation gives:

$$Q_m = C_m \cos \left(\sqrt{\Lambda_m}\, t + \phi_m \right) \quad m = 1, 2 \ldots 3N . \qquad \text{(IV, 3-9b)}$$

The mass-weighted coordinates defined in Equation (IV, 3-8b) may also be used. The solution of Newton's equation of motion is then:

$$q_j = \sum_{m=1}^{3N} b_{jm} C_m \cos \left(\sqrt{\Lambda_m}\, t + \phi_m \right) . \qquad \text{(IV, 3-10)}$$

This expression can be compared directly with Equation (IV, 3-6b). For the specific cases where only the eigenvalues λ_m are taken, the result is that:

$$q_j = \sum_{m=1}^{3N} A_{jm} \cos \left(\sqrt{\lambda_m}\, t + \varphi_m \right) \qquad \text{(IV, 3-11a)}$$

or, after introducing the amplitudes B_{jm} (see Equation (IV, 3-7)):

$$q_j = \sum_{m=1}^{3N} B_{jm} k_m \cos \left(\sqrt{\lambda_m}\, t + \varphi_m \right) . \qquad \text{(IV, 3-11b)}$$

The important conclusion which can now be made is that by comparing Equations (IV, 3-11b) and (IV, 3-10) with (IV, 3-9b), the result is obtained that $\lambda_m = \Lambda_m$. For every normal mode q_j there is one normal coordinate Q_j, and the frequency of oscillation is identical. Furthermore, it follows that $B_{jm} = b_{jm}$. The coefficients B_{jm} determine the normal modes while the b_{jm}'s are the constants which relate the normal coordinates to the normal mode q_j (see Equation (IV, 3-8b)). Thus Equation (IV, 3-8b) can also be written as:

$$q_j = \sum_{m=1}^{3N} B_{jm} Q_m \qquad \text{(IV, 3-12)}$$

or

$$\dot{q}_j = \sum_{m=1}^{3N} B_{jm} \dot{Q}_m \qquad \text{(IV, 3-13)}$$

and

$$\sum_{j}^{3N} \dot{q}_j^2 = \sum_{j,m,n} B_{jm}\dot{Q}_m B_{jn}\dot{Q}_n \qquad \text{(IV, 3-14)}$$

but (see Equation (IV, 3-7b))

$$\sum_{j} B_{jm}B_{jm} = 1 .$$

Thus, if:

$$\sum_{j} B_{jm}B_{jn} = \delta_{mn} \qquad \text{(IV, 3-15a)}$$

then

$$\sum_{j} \dot{q}_j^2 = \sum_{m} \dot{Q}_m^2 . \qquad \text{(IV, 3-15b)}$$

The transformation represented by Equations (IV, 3-15b) is an orthogonal transformation. It is also possible to arrive at the same orthogonal transformation if the appropriate expression of Q_m in terms of q_j is taken.

The relation of the a coefficients of Equation (IV, 3-8a) and the b coefficients of Equation (IV, 3-8b) can also be established. As a result of the fact that $B_{jm} = b_{jm}$, the expression (IV, 3-8c), that is:

$$\sum_{j=1}^{3N} a_{mj}b_{jn} = \delta_{mn}$$

can be rewritten as:

$$\sum_{j=1}^{3N} a_{mj}B_{jn} = \delta_{mn} . \qquad \text{(IV, 3-15)}$$

The linear transformation $q_j = \sum_{m=1}^{3N} b_{jm}Q_m$ is called the inverse of the transformation $Q_m = \sum_{j=1}^{3N} a_{mj}q_j$. From Equation (IV, 3-15), however, it is found that $a_{mj} = B_{jm}$ and thus $a_{mj} = b_{jm} = B_{jm}$. A table of the transformation coefficients, such as Table IV-4, reflects this equality of a_{jm}, b_{jm} and B_{jm}'s. Thus, in order to get the Q_m's in terms of the q_j's, the table is read along vertical lines, and the relation of q_j's in terms of the Q_j's is obtained by reading the rows from left to right.

The normal coordinates are the proper coordinates to be used in solving the Schrödinger wave equation for a system of vibrating atoms in the molecule. The total

TABLE IV-4

The transformation coefficients B_{jm} for the orthogonal transformation of q_j and Q_m.

	Q_1	Q_2	Q_3	–	–	–	Q_{3N}
q_1	B_{11}	B_{12}	B_{13}	–	–	–	B_{13N}
q_2	B_{21}	B_{22}	B_{23}	–	–	–	B_{23N}
q_3	B_{31}	B_{32}	B_{33}	–	–	–	B_{33N}
|	|						
q_{3N}	B_{3N1}	B_{3N2}	B_{3N3}	–	–	–	B_{3N3N}

vibrational wave function can be denoted by φ_v and there are $3N$-6 separate wave functions $\varphi(Q_m)$ which are related to the $3N$-6 normal coordinates Q_m. The wave equation for φ_v is:

$$-\frac{h^2}{8\pi^2}\sum_{m=1}^{3N-6}\frac{\partial^2\varphi_v}{\partial Q_m^2}+\tfrac{1}{2}\sum_{m=1}^{3N-6}\lambda_m Q_m^2\varphi_v = W_v\varphi_v. \qquad \text{(IV, 3-16)}$$

The wave equation φ_v in terms of the $3N$-6 separate functions $\varphi(Q_m)$ is then:

$$\varphi_v = \varphi(\mathbf{Q}_1)\,\varphi(\mathbf{Q}_2)\ldots\varphi(\mathbf{Q}_m) \qquad \text{(IV, 3-17)}$$

or

$$\varphi_v = \prod_{m=1}^{3N-6}\varphi(\mathbf{Q}_m).$$

Substitution of Equation (IV, 3-17) in (IV, 3-16) and dividing both sides of the equality by the product of all functions except $\varphi(\mathbf{Q}_m)$ yields the result:

$$-\frac{h^2}{8\pi^2}\frac{\mathrm{d}^2\varphi(\mathbf{Q}_m)}{\mathrm{d}Q_m^2}+\tfrac{1}{2}\lambda_r Q_n^2\varphi(\mathbf{Q}_m) = W_m\varphi(\mathbf{Q}_m). \qquad \text{(IV, 3-18a)}$$

The energy W_n is then:

$$E_{v_m} = (v_m + \tfrac{1}{2})\,h\nu_m \qquad \text{(IV, 3-18b)}$$

and the vibrational quantum number $v_m = 0, 1, 2\ldots$ etc. where ν_m is the vibrational frequency and the wave function $\varphi(Q_m)$ is:

$$\varphi(Q_m)_{v_m} = N_{v_m}\,e^{-1/2\alpha Q_m{}^2}\,H_{v_m}(\sqrt{\alpha}\cdot Q_m). \qquad \text{(IV, 3-18c)}$$

The normalization constant N_{v_n} can be written as:

$$N_{v_m} = \left[\left(\frac{\alpha}{\pi}\right)^{1/2}\frac{1}{2_{v_m}(v_m!)}\right]^{1/2} \qquad \text{(IV, 3-18d)}$$

where $\alpha = 4\pi^2\nu_m/h$ and $H_{v_m}(\sqrt{\alpha}\cdot Q_m)$ are the Hermite polynomials of degree ν_m in the normal coordinates Q_m. Some of these polynomials are given in Table IV-5. The vibrational wave functions are properly normalized, thus $\langle\varphi_{v_m}(\mathbf{Q}_m)\,|\,\varphi_{v'_m}(\mathbf{Q}_m)\rangle = \delta_{v_m v'_m}$.

TABLE IV-5

Some normalized Hermite Polynomials

$H_0 = 1$
$H_1 = 2\sqrt{\alpha}\,Q_m$
$H_2 = 4\alpha\,Q^2{}_m - 2$
$H_3 = 8\alpha\sqrt{\alpha}\,Q^3{}_m - 12\sqrt{\alpha}\,Q$
$H_4 = 16\alpha^2\,Q^4{}_m - 48\alpha\,Q^2 + 12$

where $\alpha = \dfrac{4\pi^2\nu_m}{h}$ and ν_m the frequency of oscillation.

The values of other integrals are given in Table IV-6. The Hermite polynomials are of interest in that they determine the symmetry of the vibrational wave functions. This can be seen from Equation (IV, 3-18c). Neither the normalization constant N_{v_n} nor the term $e^{-1/2\alpha Q_m^2}$ is of importance if the wave function is submitted to the symmetry operations of the group. Q_n can go into itself or into minus itself, but Q_n^2 is always invariant. If all the 3N-6 normal modes are in the quantum state $v_m=0$, (where $m=1, 2 \ldots 3N\text{-}6$), then the symmetry of the total vibrational wave function Φ_{v_m} is equal to the totally symmetric representation of a point group. This follows from the fact that the Hermite polynomials $H_{v_m=0}(\sqrt{\alpha}\cdot Q_m)$ are all equal to unity.

Wave functions for a degenerate vibrational level are derived in the following way. Suppose that the normal modes Q_A and Q_B have both a frequency of oscillation of ν. The quantum mechanical energy levels of states in which these normal modes are excited by one or more vibrational quantum numbers are also degenerate. For a non-degenerate normal mode, the energy of a level was equal to $(v+\frac{1}{2})h\nu_m$ but for the situation above, the energy is determined by two quantum numbers v_A and v_B. Thus:

$$W_V = (V+1)\,h\nu \quad \text{where} \quad V = v_A + v_B\,. \qquad \text{(IV, 3-19)}$$

For the ground state, both quantum number v_A and v_B are equal to zero, and the wave function is equal to $\psi_{00}(\mathbf{Q}_A, \mathbf{Q}_B) = \varphi(\mathbf{Q}_A)_0\,\varphi(\mathbf{Q}_B)_0$. This wave function belongs again to the totally symmetric species of the point group of the molecule. The states for which $v_A=1$ and $v_B=0$, and $v_A=0$ but $v_B=1$ have the same energy, but the wave functions are:

$$\psi_{10}(\mathbf{Q}_A, \mathbf{Q}_B) = \varphi_0(\mathbf{Q}_A)\,\varphi_0(\mathbf{Q}_B)$$

and

$$\psi_{01}(\mathbf{Q}_A, \mathbf{Q}_B) = \varphi_0(\mathbf{Q}_A)\,\varphi_1(\mathbf{Q}_B)\,. \qquad \text{(IV, 3-20)}$$

The wave functions ψ_{10} and ψ_{01} are not equal to each other. Both $\mathbf{Q}_A$ and $\mathbf{Q}_B$ transform according to a degenerate species of the point group. If $V=2$, then the following combinations are possible: $v_A=2$, $v_B=0$, $v_A=1$, $v_B=1$ and $v_A=0$, $v_B=2$. The level is three-fold degenerate and the symmetry can be determined from the functions $\mathbf{Q}_A^2$, $\mathbf{Q}_A\mathbf{Q}_B$ and $\mathbf{Q}_B^2$. For $V=3$, the level shall be four-fold degenerate. Thus, in general, the level for $V=9$ is 9+1-fold degenerate, and the wave functions can be constructed as indicated above for the simple case where the degeneracy is two.

IV-4. Theory for the Vibrational Raman Effect

In molecular vibrational Raman scattering, a transition is induced between a vibrational level v_m of a normal coordinate $\mathbf{Q}_m$ associated with the electronic state k, and another vibrational level v_m^1 of the same electronic state k. This electronic state is nearly always the electronic ground state. In letting the electronic quantum number k equals n in Equation (IV, 1-12, 13), the following expression is obtained for an element of the scattering tensor $(\alpha_{\varrho\sigma})_{k,\, v_m;\, k,\, v'_m}$, which may also be written as $(\alpha_{\varrho\sigma})_{v_m,\, v'_m}$.

$$(\alpha_{\varrho\sigma})_{v_m,\, v'_m} = \frac{e^2}{h} \sum_r \left[\frac{(M_\sigma^0)_{kr} (M_\varrho^0)_{rk}}{\bar{\nu}_{rk} - \nu_0} + \frac{[\varrho \leftrightarrow \sigma]}{\nu_{rk} + \nu_0} \right] \langle \varphi_{v_m}(\mathbf{Q}_m) \mid \varphi_{v'_m}(\mathbf{Q}_m) \rangle +$$

$$+ \frac{e^2}{h} \sum_r \left\{ \sum_t \frac{\{(M_\sigma^0)_{kr} (M_\varrho^0)_{rt} + (M_\varrho^0)_{rk} (M_\sigma^0)_{tr}\} h_{tk}^{Q_m} + \{(M_\sigma^0)_{kr} (M_\varrho^0)_{kt} + (M_\varrho^0)_{rk} (M_\sigma^0)_{kt}\} h_{tr}^{Q_m}}{\bar{\nu}_{r,k} + \nu_0} \times \right.$$

$$\left. \times \langle \varphi_{v_m}(\mathbf{Q}_m) | \, Q_m \, | \varphi_{v'_m}(\mathbf{Q}_m) \rangle \right\} + \frac{e^2}{h} \sum_r \left[\sum_{t,\, Q} \frac{[\varrho \leftrightarrow \sigma]}{\bar{\nu}_{rk} + \nu_0} \right] \times$$

$$\times \langle \varphi_{v_m}(\mathbf{Q}_m) | \, \mathbf{Q}_m \, | \varphi_{v'_m}(\mathbf{Q}_m) \rangle + \cdots. \qquad \text{(IV, 4-1)}$$

The results given here are similar to those derived by Albrecht, who directly employed vibronic wave function in the expression for the scattering tensor $(\alpha_{\varrho\sigma})_{k,\, v_m;\, k,\, v'_m}$.

For a molecule, there are 3N-6 normal modes of vibration, and the expression derived in the above in only valid in the case where the vibrational and rotational states of the molecules do not interact. The first member of Equation (II, 4-1) can be different from zero if the vibrational quantum number $v'_m = v_m$, and the second and third members can be different from zero if the matrix element $\langle \varphi_{v_m}(\mathbf{Q}_m) | \, \mathbf{Q}_m \, | \varphi_{v'_m}(\mathbf{Q}_m) \rangle \neq 0$. For pure harmonic oscillator type wave functions, this matrix element is non-vanishing if $v'_m = v_m \pm 1$. The condition $v'_m = v_m$ implies that the first member contributes to Rayleigh scattering, and the second and third members represent vibrational Raman effects because the emitted radiation of the moment $\mathbf{M}^1_{kv_m;\, kv'_m}$ is according to (I, 4-23b) and (I, 4-25) equal to $\nu_0 \pm \nu_{v'_m,\, v'_m}$. At the basis of the above expressions for the vibrational Raman scattering of molecules lies the concept of vibronic coupling. The important approximation made is that of Equation (IV, 1-9a), where the Hamiltonian of the molecule is expanded in a Taylor series:

$$\mathscr{H}(\mathbf{r}, \mathbf{Q}_m) = \mathscr{H}(\mathbf{r}, \mathbf{Q}_m = 0) + \left(\frac{\partial \mathscr{H}}{\partial Q_m} \right)_{Q_m = 0} \mathbf{Q}_m + \cdots. \qquad \text{(IV, 4-2)}$$

The value of the tensor elements $(\alpha_{\varrho\sigma})_{v_m,\, v'_m}$ depends on the integral $\langle v_m | \, \mathbf{Q} \, | v_m \rangle$, and also on the values of the matrix elements $(M_\sigma^0)_{kr}$ and the constants $h_{tk}^{Q_m}$ etc. For the vibrational Raman transition $v'_m = 1 \leftarrow v_m = 0$, the lowest state of the system is the ground state k, $v_m = 0$. The states t and r of Equation (IV, 4-1) are electronic excited states of the molecule, and they belong to the irreducible representation of the point group of the molecule. Among the large number of states t and r, there is always some combination to be found so that the totally symmetric representation occurs in the product $\Gamma_r \times \Gamma_{\varrho\sigma} \times \Gamma_t$. It is thus necessary that conditions should be found for which $(M_r^0)_{kr}$, $(M_r^0)_{kt}$ and $h_{tr}^{Q_m}$ are simultaneously different from zero. The matrix elements of the electric dipole operator are non-vanishing if the excited states r and t belong to representations to which the coordinates x, y or z also belong. In order that $h_{tk}^{Q_m} =$

$=\langle\psi_t^0|\,(\partial\mathscr{H}/\partial Q_m)_{Q_m=0}\,|\psi_k^0\rangle\neq 0$, it is necessary to know the transformation properties of $\partial\mathscr{H}/\partial Q_m$. These properties follow directly from Equation (II, 4-2). The Hamiltonian of the molecule is always an invariant, and $\Gamma_{\partial\mathscr{H}/\partial Q_m}\equiv\Gamma_{Q_m}$. Thus, $h_{tr}^{Q_m}\neq 0$ if $\Gamma_{\psi^0_t}\times\Gamma_{Q_m}\times\Gamma_{\psi^0_r}$ contain the totally symmetric species, and thus Γ_{Q_m} must correspond to at least one of the species contained in the direct product $\Gamma_{\psi^0_t}\times\Gamma_{\psi^0_r}$. These species are one of, or a combination of Γ_{x^2}, Γ_{xy}, $\Gamma_{xz}-\ldots\,\Gamma_{z^2}$ and they also represent the transformation properties of the elements of the scattering tensor. It is of interest to note that – according to the above – electronic wave functions of the molecule play an important role in the intensity of the vibrational Raman lines.

In chapter I, the difference between the scattering and polarizability tensor was explained. The elements $(\alpha_{\varrho\sigma})_{v_m,\,v'_m}$ belong to the former, and the electronic polarizability tensor has the elements $(\alpha_{\varrho\sigma})_{kk}$. An expression for the elements of the scattering tensor may be obtained in Equation (IV, 1-2) by replacing the electronic quantum number n by k so that:

$$(\alpha_{\varrho\sigma})_{k,\,v_m;\,k,\,v'_m}=$$
$$=\frac{e^2}{h}\sum_{r,\,v''_m}\left[\frac{\langle\psi_k^0\cdot\varphi_{v_m}(Q_m)|\,\sigma\,|\psi_r^0\cdot\varphi_{v''_m}(Q_m)\rangle\,\langle\psi_r^0\cdot\varphi_{v''_m}(Q_m)|\,\varrho\,|\psi_k^0\cdot\varphi_{v'_m}(Q_m)\rangle}{\nu_{rv''_m,\,kv_m}-\nu_0}+\right.$$
$$\left.+\frac{[\varrho\rightarrow\sigma]}{\nu_{rv''_m,\,kv'_m}+\nu_0}\right].\qquad\text{(IV, 4-3a)}$$

In the case where the sum rule can be applied, the result is:

$$(\alpha_{\varrho\sigma})_{k,v;k,v'}=\frac{1}{h}\sum_r\frac{\langle\psi_k^0\varphi_{v_m}(\mathbf{Q}_m)|\,e\sigma\,|\psi_r^0\rangle\,\langle\psi_r^0|\,e\varrho\,|\psi_k^0\varphi_{v'_m}(Q_m)\rangle}{\bar{\nu}_{rk}-\nu_0}+$$
$$\left.+\frac{[\varrho\leftrightarrow\sigma]}{\bar{\nu}_{rk}+\nu_0}\right].\qquad\text{(IV, 4-3b)}$$

The electric dipole operator acts only on the electronic wave function so that

$$(\alpha_{\varrho\sigma})_{k,\,v_m;k,v'_m}=\langle\varphi_{v'_m}(\mathbf{Q}_m)|\,(\alpha_{\varrho\sigma})_{kk}\,|\varphi_{v_m}(\mathbf{Q}_m)\rangle\qquad\text{(IV, 4-4a)}$$

and the elements of the polarizability tensor are given by:

$$(\alpha_{\varrho\sigma})_{kk}=\frac{1}{h}\sum_r\left[\frac{\langle\psi_k^0|\,e\sigma\,|\psi_r^0\rangle\,\langle\psi_r^0|\,e\varrho\,|\psi_k^0\rangle}{\bar{\nu}_{rk}-\nu_0}+\frac{[\varrho\leftrightarrow\sigma]}{\bar{\nu}_{rk}+\nu_0}\right].\qquad\text{(IV, 4-4b)}$$

At the basis of Placzek's polarizability theory for the vibrational Raman effect lies the expansion of the electronic polarizability α_{kk} in a Taylor series to the normal coordinates. Thus:

$$\alpha_{kk}=\alpha^0+\left(\frac{\partial\alpha_{kk}}{\partial Q_m}\right)_{Q_m=0}\cdot Q_m+\frac{1}{2}\left(\frac{\partial^2_{\alpha_{kk}}}{\partial Q_m^2}\right)_{Q_m=0}Q_m^2+\cdots.\qquad\text{(IV, 4-5a)}$$

The zero point around which the expansion is carried out is again fixed at the equilibrium position of the nuclei of the atoms of the molecule. Substitution of Equation (IV, 4-5) in the expression (IV, 4-4a) leads to:

$$\langle \varphi_{v'_m}(\mathbf{Q}_m) \mid (\alpha_{\varrho\sigma})_{kk} \mid \varphi_{v_m}(\mathbf{Q}_m)\rangle = \alpha_0 \langle \varphi_{v'_m}(\mathbf{Q}_m) \mid \varphi_{v_m}(\mathbf{Q}_m)\rangle + + \left(\frac{\partial(\alpha_{\varrho\sigma})_{kk}}{\partial Q_m}\right) \langle \varphi_{v'_m}(\mathbf{Q}_m) \mid (\mathbf{Q}_m) \mid \varphi_{v_m}(\mathbf{Q})\rangle + + \frac{1}{2}\left(\frac{\partial^2(\alpha_{\varrho\sigma})_{kk}}{\partial Q_m^2}\right)_{Q_m=0} \langle \varphi_{v'_m}(\mathbf{Q}_m) \mid \mathbf{Q}_m^2 \mid \varphi_{v_m}(\mathbf{Q}_m)\rangle + \cdots. \qquad \text{(IV, 4-5b)}$$

The first member is different from zero if $v^1 = v$ and the frequency of the emitted light is equal to the frequency of the incident radiation.

The integral $\langle \varphi_{v'_m}(\mathbf{Q}_m) | \mathbf{Q}_m | \varphi_{v_m}(\mathbf{Q}_m)\rangle$ for $v'_m = v_m \pm 1$ and it follows that the frequency of the radiation field is equal to $\nu = \nu_0 + | \nu_m |$. Thus the position of bands of Stokes and anti-Stokes Raman spectrum, relative to the frequency ν_0, is a direct measure of the frequency of oscillation of the normal modes of the molecule. The fundamental vibrations can only be observed if

$$\left(\frac{\partial(\alpha_{\varrho\sigma})_{kk}}{\partial Q_m}\right)_{Q_m=0} \neq 0 .$$

The electronic polarizability α_{kk}, given by Equation (IV, 4-4) is to first approximation independent of the motions of the atoms of the molecule. During the vibrational motion α_{kk} changes and the dependence of α on the normal coordinate $\mathbf{Q}_m$ is designated by $\alpha_{kk}^{Q_m}$, and Equation (IV, 4-4a) should thus read:

$$(\alpha_{\varrho\sigma})_{k,\, v_m;\, kv'_m} = \langle \varphi_{v'_m}(\mathbf{Q}_m) | (\hat{\alpha}_{\varrho\sigma}^{Q_m})_{kk} | \varphi_{v_m}(\mathbf{Q}_m)\rangle . \qquad \text{(IV, 4-6a)}$$

It follows from (IV, 4-5a) that:

$$(\hat{\alpha}_{\varrho\sigma}^{Q_m})_{kk} = \left(\frac{\partial(\alpha_{\varrho\sigma})_{kk}}{\partial Q_m}\right)_{Q_m=0} \cdot \mathbf{Q}_m . \qquad \text{(IV, 4-6b)}$$

Also, for $v'_m - v_m \pm 1$ (from (IV, 4-6b)):

$$\Gamma_{(\hat{\alpha}Q_m/\varrho\sigma)kk} = \Gamma_{Q_m} \qquad \text{(IV, 4-6c)}$$

and (from (IV, 4-5)):

$$\Gamma_{(\partial(\hat{\alpha}_{\varrho\sigma})_{kk}/\partial Q_m)Q_m=0} = \Gamma_{Q_m} . \qquad \text{(IV, 4-6d)}$$

The elements of the polarizability tensor α_{kk} transform according to the totally symmetric species of the point group of the molecule. These tensor elements are of course directly related to the principal values of polarizability and the axis of highest symmetry of the polarization ellipsoïde coincides with the axis of highest symmetry of the molecule.

The polarizability tensors is thus characterized by the elements of $(\alpha_{\varrho\varrho})_{kk}$ and all $(\alpha_{\varrho\sigma})_{kk} = 0$. In the Appendix of this book are given character tables and transformation properties of the tensors $\alpha_{\varrho\sigma}$, α_Q^k and $\beta_{\varrho\sigma\sigma'}$. The transformation properties are those of a general tensor α (of rank two) and β (of rank three). The intensity of a vibrational Raman band for the transition $v'_m=1 \leftrightarrow v_m=0$ (or in general $\Delta v_m=\pm 1$) depends on the value of $(\alpha'_{\varrho\sigma})_{kk}=(\partial(\alpha_{\varrho\sigma})_{kk}/\partial Q_m)_{Q_m=0}$ and only if $\Gamma_{Q_m}=\Gamma_{(\alpha'\varrho_\sigma)kk}$ becomes the transition Raman active. In arriving at the selection rules the symmetry properties of the derivative of the elements of the polarizability tensor (at $Q_m=0$) with respect to the normal coordinate $\mathbf{Q}_m$ replaces thus the symmetry properties of the elements of the polarizability tensor and it should be emphasized that if an element is zero, this in general is not so for the derivative.

The selection rules of the vibrational Raman effect may also be derived by inspection of the behaviour of the polarizability tensor if the molecule vibrates.

The principal values of polarizability of a molecule of C_{2v} symmetry are $\alpha_{xx} \neq \neq \alpha_{yy} \neq \alpha_{zz}$. This follows from the invariance of α. The principal vectors are the x, y and z-axis of the coordinate system shown in Figure, IV-2, and the semi-axis of the ellipsoid are all different (see Equation (II, 2-8) and (II, 2-9)).

During the vibrational motion of the A_1 modes, it is seen that the polarization ellipsoid contracts and expands along the principal vectors. Thus the direction of the semi-axis does not change and $(\partial\alpha_{xx}/\partial Q_1)_{Q_1=0} \neq 0$, $(\partial\alpha_{yy}/\partial Q_1)_{Q_1=0} \neq 0$,$(\partial\alpha_{zz}/\partial Q_1)_{Q_1=0} = =0$, also in this case $\alpha_{xx} \neq 0$, $\alpha_{yy} \neq 0$, $\alpha_{zz} \neq 0$. Thus the A_1 modes are Raman active. On the other hand, for the B_1 modes, the value of the component $\alpha_{xz}=0$ (for $Q_2=0$). But during the vibrational motion, away from the equilibrium position, the polarization ellipsoid becomes distorted to such a degree, that the direction of the principal vectors do not coincide with the x, y- and z-axis of Fig. IV-2. The deviation reaches a maximum if $Q=Q$max. It is always possible to take the x-, y- and z-axis as the principal vectors, but a tensor having only diagonal elements if the proper eigenvectors are used is carried over in a tensor having diagonal as well as off-diagonal elements if the eigenvectors are the x-, y- and z-axis. This follows from Equation (II, 2-5). The result in this case is that $(\partial\alpha_{zx}/\partial Q_2)_{Q_2=0} \neq 0$.

For the three normal modes of vibration of a linear molecule shown in Figure IV-3,

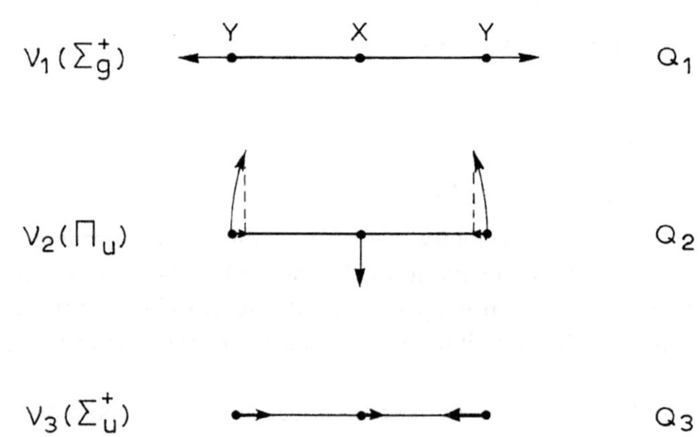

Fig. IV-3. The three normal modes of a linear triatomic molecule (XY_2) of symmetry $D\infty h$. The ν_2 mode is degenerate and only one of the normal coordinates is shown.

the Raman activity is visually derived in the following way. The polarization ellipsoid contracts and expands along the semi-axis during the vibrational movements of all three normal modes and consequently the direction of the principal vectors are not altered. Thus only $(\partial\alpha_{\varrho\varrho}/\partial Q_1)_{Q_1=0} \neq 0$ and the ν_1 mode is the only one active in the Raman spectrum. Although at first glance there seems to be a difference between the theory of Placzek and Albrecht this is not really so. The polarizability tensor is related to electronic wave functions of the system and their dependence on the normal modes is achieved in a indirect way (via the dependence of α), whilst in Albrechts vibronic coupling model the dependence is obtained in a direct way. The vibronic wave functions are later on used to obtain an expression for an element of the scattering tensor. The matrix element $\langle\varphi(\mathbf{Q})_{v_m+1}|\,(\mathbf{Q}_m)_{kk}\,|\varphi(\mathbf{Q}_m)v_m\rangle = ((v_m+1)/2\alpha)^{1/2}$ where $\alpha = 4\pi^2\nu_m/h$ – see Table IV-6 and the intensity of the transition $I_{v'_m, v_m+1}$ is thus proportional to v_m+1.

TABLE IV-6

Values of some important matrix elements

$$\langle\varphi_{v'_m}(\mathbf{Q}_m)\mid\varphi_{v_m}(\mathbf{Q}_m)\rangle = \begin{cases} 0 & \text{if } v'_m \neq v_m \\ 1 & \text{if } v_m = v_{m'} \end{cases}$$

$$\langle\varphi_{v'_m}(\mathbf{Q}_m)\mid Q_m\mid\varphi_{v_m}(\mathbf{Q}_m)\rangle = \begin{cases} 0 & \text{if } v'_m \neq v_m \pm 1 \\ \left(\dfrac{v_m}{2\alpha}\right)^{1/2} & \text{if } v'_m = v_m - 1 \\ \left(\dfrac{v_m+1}{2\alpha}\right)^{1/2} & \text{if } v'_m = v_m + 1. \end{cases}$$

$$\langle\varphi_{v'_m}(\mathbf{Q}_m)\mid \mathbf{Q}_m{}^2\mid\varphi_{v_m}(\mathbf{Q}_m)\rangle = \begin{cases} 0 & \text{if } v'_m \neq v_m \text{ or } v'_m = v_m \pm 2 \\ \dfrac{1}{2\alpha}\,[(v_m+1)(v_m+2)]^{1/2} & \text{if } v'_m = v_m + 2 \\ \dfrac{(v_m+\frac{1}{2})}{\alpha} & \text{if } v'_m = v_m \\ \dfrac{1}{2\alpha}\,[v_m(v_m-1)]^{1/2} & \text{if } v'_m = v_m - 2. \end{cases}$$

$$\langle\varphi_{v'_m}(\mathbf{Q}_m)\mid \mathbf{Q}_m{}^3\mid\varphi_{v_m}(\mathbf{Q}_m)\rangle = \begin{cases} 0 & \text{if } v'_m \neq v_m \pm 1 \text{ or } v'_m \neq v_m \pm 3 \\ \left[\dfrac{(v_m+1)(v_m+2)(v_m+3)}{8\alpha^3}\right]^{1/2} & \text{if } v'_m = v_m + 3 \\ 3\left[\dfrac{(v_m+1)^3}{8\alpha^3}\right]^{1/2} & \text{if } v'_m = v_m + 1. \\ 3\left[\dfrac{v_m{}^3}{8\alpha^3}\right]^{1/2} & \text{if } v'_m = v_m - 1 \\ \left[\dfrac{v_m(v_m-1)(v_m-2)}{8\alpha^3}\right]^{1/2} & \text{if } v'_m = v_m - 3 \end{cases}$$

If N_v is the number of particles which occupies the vth vibrational level, then:

$$I_{v_m, v'_m} = \frac{16\pi^4(\nu_0 + \nu_{v'_m, v_m})^4}{3c^3} N_{v_m} \frac{v_m+1}{2\alpha}\left(\frac{\partial\alpha_{kk}}{\partial Q_m}\right)^2_{Q_m=0}. \tag{IV, 4-7}$$

At temperatures >0 K, excited vibrational states are populated and the total number of particles N_T is thus equal to

$$N_T = N_0 + N_1 + N_2 + \cdots. \tag{IV, 4-8a}$$

If the Boltzmann distribution law holds, then:

$$\frac{N_{v+1}}{N_v} = e^{-h\nu_m/kT}$$

so that:

$$N_T = N_0 \left(1 + e^{-h\nu_m/kT} + e^{-2h\nu_m/kT} + \ldots\right) = \frac{N_0}{1 - e^{-h\nu_m/kT}}. \qquad \text{(IV, 4-8b)}$$

In principle, the intensity of a Stokes Raman band of the spectrum is made up of the transitions $I_{01} + I_{12} + I_{23} + \cdots$ and for harmonic oscillator type wave functions, the individual transitions all have the same energy. Thus, for Stokes radiation:

$$I_T = N_0 I_{01} + N_1 I_{12} + N_2 I_{23} + \cdots$$

or

$$\begin{aligned} I_T &= N_0 I_{01} \left[1 + 2e^{-h\nu_m/kT} + 3e^{-2h\nu_m/kT} + \ldots\right] \\ &= N_0 I_{01} \left[(1 - e^{-h\nu_m/kT})^{-1} + e^{-h\nu_m/kT}(1 - e^{-h\nu_m/kT})^{-2}\right] \\ &= N_T I_{01} (1 - e^{-h\nu_m/kT})^{-1}. \end{aligned} \qquad \text{(IV, 4-9)}$$

For the intensity of the anti-Stokes band:

$$I_T = N_T I_{10} (e^{+h\nu_m/kT} - 1)^{-1} \qquad \text{(IV, 4-10)}$$

and the ratio of the intensity of Stokes and anti-Stokes radiation is equal to

$$\frac{I_{\text{St}}}{I_{\text{a-St}}} = \left(\frac{\nu_0 - \nu_m}{\nu_0 + \nu_m}\right)^4 e^{h\nu_m/kT}.$$

The third member of Equation (IV, 4-5b) is different from zero if $(\partial^2(\alpha_{\varrho\sigma})_{kk}/\partial Q_m^2) \neq \neq 0_{Q_m=0}$ and $v_m' = v_m$ and $v_m' = v_m + 2$. From the frequency of the corresponding induced moments, it follows that this term contributes to Rayleigh* and Raman scattering. The Raman transition is the first overtone because the molecule experiences two jumps in the vibrational quantum number. The Raman selection rule for the transition depends on the integral $\langle \varphi(\mathbf{Q}_m)_{v_m+2} | (\alpha_{\varrho\sigma})_{kk} | \varphi(\mathbf{Q}_m)_{v_m} \rangle$. For all non-degenerate normal coordinates, the wave function $\varphi(\mathbf{Q}_m)_{v_m+2}$ belongs to the same species as the function $\varphi(Q_m)_{v_m}$, and consequently, the tensors associated with overtones of non-degenerate modes belong to the totally symmetric species of the point group of the molecule. The temperature dependence of an overtone is $I_T = N_T I_{02} (1 - e^{-h\nu_m/hT})^{-2}$. The expansion given in Equation (IV, 4-5) may also be carried out as follows:

$$\alpha = \alpha^0 + \left(\frac{\partial \alpha_{kk}}{\partial Q_A}\right)_{Q_A=0} \mathbf{Q}_A + \tfrac{1}{2}\left(\frac{\partial^2 \alpha_{kk}}{\partial Q_A \partial Q_B}\right)_{Q_A=Q_B=0} \mathbf{Q}_A \mathbf{Q}_B + \ldots. \qquad \text{(IV, 4-11)}$$

Here, $\mathbf{Q}_A$ and $\mathbf{Q}_B$ are different normal modes of the molecule. In order to find the selection rules for a combination band, the total vibrational wave function should be considered. It follows that the transition associated with the third member is an overtone where both vibrational modes are excited by one vibrational quantum, and

* This is incoherent Rayleigh scattering.

the overtone is only Raman active if elements of the Raman tensor belong to species to which the normal coordinates $\mathbf{Q}_A$ and $\mathbf{Q}_B$ also belong. Overtones can also be found where one normal coordinate is excited by two or more quanta, but it may also be possible that one mode is excited by n quanta and other modes are de-excited by m quanta (combination). Overtones and combination bands may also involve degenerate vibrational modes. In Table IV-7 are given the species of some vibrational levels of a degenerate normal mode of a molecule which belongs to the point

TABLE IV-7

Species of levels of normal coordinates of molecules of symmetry D_{3h}

Symmetry of normal coordinate	Vibrational quantum number $v_m = 0$	$v_m = 1$	$v_m = 2$	$v_m = 3$
A'_1	A'_1	A'_1	A'_1	A'_1
A'_2	A'_1	A'_2	A'_1	A'_2
E'	A'_1	E'	$A'_1 + E'$	$A'_1 + 2E'$
A''_1	A'_1	A''_1	A''_1	A''_1
A''_2	A'_1	A''_2	A''_1	A''_2
E''	A'_1	E''	$A''_1 + E''$	$A''_1 + 2E''$

group D_{3h}. It is seen that in the product of the irreducible representation of $v_m=0$ and $v=1$ states, only the species E' occurs. But the species occurring in the product of representations to which the wave functions of the $v_m=1$ and $v_m=2$ states belong are: $E' \times (A'_1+E') = E'+E'+A'_1$. The species A'_1 is not present in the product of representations of ground state and first excited state, and it appears that the Raman tensor linking $v_m=1$ and $v_m=0$ is not equal to the tensor of $v_m=2$ and $v_m=1$. That the same tensor also links successive levels of the degenerate normal modes of vibration can be shown as follows. If the wave functions of the degenerate mode are denoted with $\varphi(\mathbf{Q}_A)_v$ and $\varphi(\mathbf{Q}_B)_v$, then for the first excited level, the wave functions are $\varphi(\mathbf{Q}_A)_1$ and $\varphi(\mathbf{Q}_B)_1$. Both wave functions belong to the species E' of the group because $\mathbf{Q}_A$ and $\mathbf{Q}_B$ transform similarly to E'. For the second excited site, the wave functions are $\varphi(\mathbf{Q}_A)_2 \cdot \varphi(\mathbf{Q}_B)_0$; $\varphi(\mathbf{Q}_A)_{v=1}\ \varphi(\mathbf{Q}_B)_{v=1}$ and $\varphi(\mathbf{Q}_A)_0 \cdot \varphi(\mathbf{Q}_B)_2$. The transformation properties of these three functions are also represented by the normal coordinates $\mathbf{Q}_A^2$, $\mathbf{Q}_A \cdot \mathbf{Q}_B$ and $\mathbf{Q}_B^2$ respectively. If these three functions are subjected to the symmetry operations of the group D_{3h}, it is found that the combinations $\mathbf{Q}_A^2+\mathbf{Q}_B^2$ transform similarly to the totally symmetric species, and to the species E'_1 belong both $\mathbf{Q}_A^2-\mathbf{Q}_B^2$ and $\mathbf{Q}_A \cdot \mathbf{Q}_B$. These functions may now be used to calculate the matrix element $\langle v=2|\ Q\ |v=1\rangle$. Here $v=2$ represents the wave functions for the second excited state; $\mathbf{Q}$ stands for either $\mathbf{Q}_A$ or $\mathbf{Q}_B$ and $v=1$ is the wave function for the first excited state. This matrix element is only different from zero if $v=2$ is $\mathbf{Q}_A^2+\mathbf{Q}_B^2$ and $v=1$ is $\mathbf{Q}_A$ or $\mathbf{Q}_B$. The non-vanishing elements of the Raman tensor are thus of species $A'_1 \times E' = E'$, identical to the tensor for $v=1 \leftarrow v=0$.

Combination bands involving degenerate and non-degenerate normal modes can also be Raman active. In general, however, it is found that the intensity of overtones and combination bands are much weaker than the intensity of the fundamentals ($1 \leftarrow 0$ etc. transitions). Part of this decrease in intensity may be related to the fact that in Placzek's polarizability theory, the expansion has to be carried out to at least the second order. This is also the case for Albrechts vibronic model, but there the Hamiltonian should be expanded to terms in second order of the normal coordinates.

Raman spectra can become complicated as a result of the occurrence of Fermi-resonance. Here the overtone of one of the fundamental vibrations of a molecule has approximately the same frequency as – is in resonance with – another fundamental mode of the molecule. A typical example occurs in the Raman spectrum of the CS_2 molecule where resonance occurs of $2\nu_2$ and ν_1. The three normal modes are depicted in Figure IV-3 and it is seen that the ν_2 mode has a component along the internuclear axis. The overtone $2\nu_2$ has about the same frequency shift as ν_1, and the level W_2^0 (for $v=2$ of the ν_2 mode), and the level W_1^0 (for $v=1$ of the ν_1 mode) repel each other because they have the same symmetry. The actual position of the levels is such that the energy difference is $\Delta W = \pm\sqrt{(4W'^2 + (W_2^0 - W_1^0)^2}$ where W^1 is the interaction energy of the two modes. The wave functions of the levels are then:
for

$$v = 2 \quad \text{of} \quad \nu_2: \quad \varphi_2 = a_1\varphi_2^0 + a_2\varphi_1^0$$

and for

$$v = 1 \quad \text{of} \quad \nu_1: \quad \varphi_1 = a_2\varphi_2^0 - a_1\varphi_1^0 .$$

The a coefficients are related to W_1^0, W_2^0 and W^1, but as a result of admixture of wave functions, part of the intensity of the 'fundamental vibration' is carried over into that of the 'first overtone' and vice versa. Fermi-resonance may be looked upon as a special case of the Auger effect. In the latter, a vibrational or rotational level associated with one particular electronic state of the system has an energy equal to the continuum of levels associated with another electronic state of the system (above the ionization limit). As a result of this type of resonance, the vibrational or rotational level may be broadened or its line shape may be affected in a particular way. Although Fermi-resonance has been observed, up to now an Auger Raman effect has not been detected. Further complications in the interpretation of vibrational Raman spectra is the coupling of rotational and vibrational motions and the occurrence of deviations of pure harmonic oscillator type vibrational wave functions. The former shall be discussed in the following chapter. The effect of anharmonicity does not alter the selection rules for the various types of vibrational Raman bands, but changes do occur in the frequency spectrum. The effect of anharmonicity is normally treated by introducing additional terms in the potential function. The total vibrational energy for the $3N$-6 normal modes of vibration of a molecule was given by:

$$W_T = \sum_{m=1}^{3N-6} \left(v_m + \frac{d_m}{2} \right) h\nu_m \qquad \text{(IV, 4-12)}$$

where d_m is the degeneracy of the mth normal coordinate. If anharmonic terms are introduced, then the total vibrational energy is:

$$W_T = \sum_{m=1}^{3N-6} (v_m + d_m/2)\, h\nu_m + \\ + \sum_m \sum_{lm} [\chi_{ml}(v_m + d_m/2)(v_l + d_l/2) + g_{ml} l_m l_l] + \cdots . \qquad \text{(IV, 4-13)}$$

The second term is directly related to the anharmonic terms, but the third term arises only in the special case for degenerate normal coordinates.

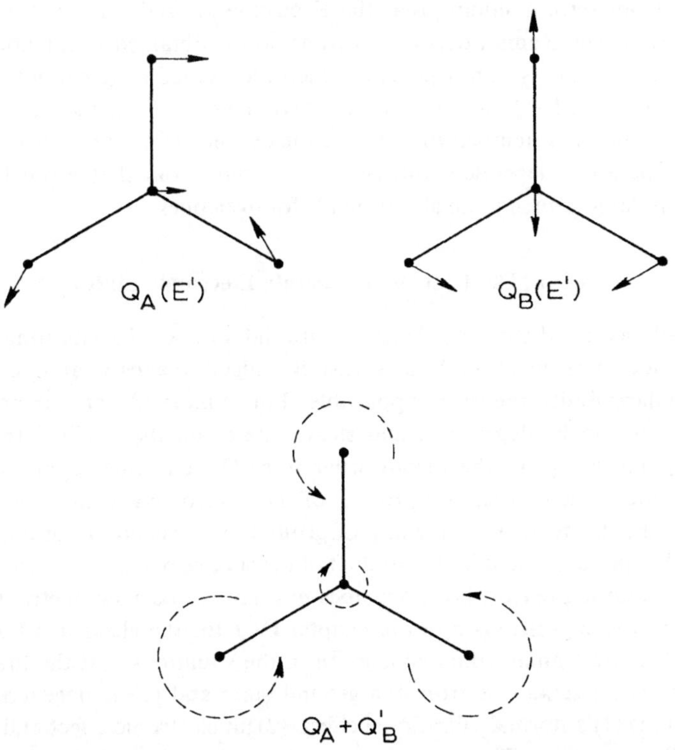

Fig. IV-4. The degenerate normal coordinates Q_A and Q_B of a molecule of symmetry D_{3h} and the approximate movement of the atoms if the phase difference of Q_B and Q_A is 90°.

From Figure, IV-4 it may be seen that a rotational motion exists if the normal coordinates Q_A and Q_B are excited so that their amplitudes are equal but out of phase by 90°. The angular momentum associated with this motion is denoted by l_m and the magnitude of the momentum is given by:

$$\frac{h}{2\pi} l_m \quad \text{where} \quad l_m = v_m, v_m - 2, \ldots 1 \quad \text{or} \quad 0 . \qquad \text{(IV, 4-14)}$$

Thus, for the degenerate mode ν_2 of the molecule of D_{3h} symmetry, the species of the levels of the second excited state were:

$$\text{for } v_m = 2 \quad \Gamma = A_1' + E' . \qquad \text{(IV, 4-15)}$$

The angular momentum is then according to Equation, (IV, 4-15):

$$l = 2, 0 .$$

The state labelled with the species A_1 has $l=0$ and the state of species E' has $l=2$. These two states can have different energy if the third member of Equation (IV, 4-13) is different from zero, assuming that the electronic ground state of the molecule is non-degenerate. The Raman tensor of a transition is obtained in the normal way. If the normal coordinate is non-degenerate, then the excited state cannot be split; but the transitions $0 \rightarrow 1$, $1 \rightarrow 2$, $2 \rightarrow 3$ etc. do not have the same frequency, and the Raman band becomes broad at temperatures where the excited states are sufficiently populated. The temperature dependent intensity is different from that given by Equation (IV, 4-9). Similar arguments can also be made for overtones.

IV-5. The Case of Degenerate Electronic States

It is normally assumed that the electronic ground state of the vibrating molecule is non-degenerate. For most molecules this is indeed the case and, consequently, Placzek's polarizability theory is applicable. For a molecule of symmetry D_{3h}, an electronic state can be degenerate and such a state can then belong to the species E' of this point group. In the zeroth order Born-Oppenheimer approximation, the total wave function is written as a product of the electronic and the vibrational wave function. Thus, the symmetry labelling of ground and excited vibrational states follows from the species present in the product of the species of electronic and vibrational states. The scattering tensor, however, does not have to be a symmetric tensor. This follows from the expressions given in chapter IV-1 for the electronic Raman effect. In Figure IV-5 are Raman transitions given in the situations that the first electronic excited state is (a) separated from the ground state and (b) is degenerate. For the former, there is (1) a normal vibrational effect, (2) an electronic effect and (3) a vibro-electronic Raman effect. The scattering tensors of the effects under (2) and (3) are, in principle, not symmetric (see Equations (IV, 1-12, 13)). Compared to the non-degenerated case, the electronic effect result in Rayleigh scattering in the degenerate situation, and the vibro-electronic Raman effect becomes a vibrational Raman effect. But the tensor for the vibro-electronic effect is not symmetric, and the overall result is that asymmetric scattering tensors can contribute to Raman intensities.

The product of representations is commonly referred to as the direct product. Such products are sufficient to determine the species of an excited state of a *degenerate* normal coordinate. For instance, in the case of a molecule of symmetry D_{3h}, the species of $v=2$ of a degenerate normal coordinate of symmetry E' is: $E' \times E' = A_1' + E'$. The normal coordinates may be designated by $\mathbf{Q}_A$ and $\mathbf{Q}_B$, thus there are two functions, f_1 and f_2,

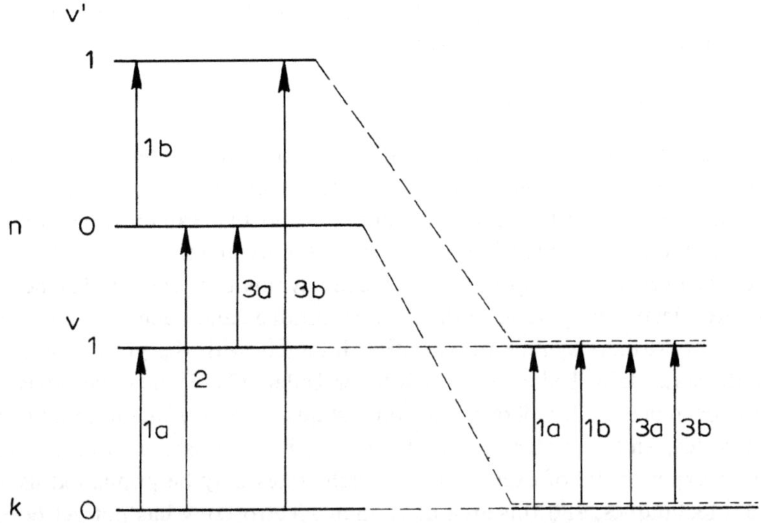

Fig. IV-5. Various Raman transitions for the situation that the grandstate becomes degenerate. n and k are electronic excited and groundstate, v' and v are vibrational quantum numbers. (1a) and (1b) – normal vibrational; (2) – electronic; (3) – vibro-electronic.

which belong to the species E' of D_{3h}. The product of the representation E' with itself is thus represented by $f^2=4$ functions of the type $f_1 \times f_2$ ($f_1^2, f_1 \cdot f_2, f_2 \cdot f_1$ and f_2^2). The product representation is generally reducible, and $f(f+1)/2$ functions of the type $f_i f_j + f_j f_i$ ($i=j$ and $i \neq j$) can belong to some reducible representation; the functions $f_i f_j - f_j f_i$ ($i \neq j$) to another. There are $f(f-1)/2$ functions of the latter type. In the example used above, the $f(f+1)/2$ functions are $\mathbf{Q}_A^2$, $\mathbf{Q}_A \cdot \mathbf{Q}_B = \mathbf{Q}_B \mathbf{Q}_A$ and $\mathbf{Q}_B^2$. They belong to a reducible representation which arises from the symmetric product of the representation E' with itself. ($E' \times E' = A_1' + E'$; $\mathbf{Q}_A^2 + \mathbf{Q}_B^2$ transforms like A_1' and $\mathbf{Q}_A \cdot \mathbf{Q}_B$ and $\mathbf{Q}_A^2 - \mathbf{Q}_B^2$ belong E'.) On the other hand, the $f(f-1)/2$ functions belong to some representation which arises from the antisymmetric product of E' with itself ($[E' \times E']$). If the functions f_1 and f_2 are normal coordinates, then the only function occurring in the anti-symmetric product is $\mathbf{Q}_A \mathbf{Q}_B - \mathbf{Q}_B \mathbf{Q}_A = 0$. This result does not only pertain to the normal coordinates of a molecule of symmetry D_{3h}, but is valid for all pertinent cases which involve degenerate vibrational levels. For electronic levels, however, the antisymmetric product has to be taken into account so that the direct product of a representation with itself is equal to the sum of the symmetric product and antisymmetric procuct. Now, $[E' \times E'] = A_2'$. Consequently, the tensor of a Raman transition between vibrational levels of the totally symmetric vibration can now be asymmetric because the selection rule follows from: $\Gamma_\alpha = E' \times E' \times A_1' + [E' \times E'] \times A_1'$ and symmetric and antisymmetric tensors belong to species occurring in the symmetric and antisymmetric products. If the normal coordinate is of species E', then the angular momentum (l) of this degenerate vibrational mode can couple with the angular momentum Λ_e of the degenerate electronic ground state to yield a vibronic angular momentum K so that:

$$K = |\pm \Lambda_e \pm l| \qquad \text{(IV, 5-1)}$$

and the result is that the degeneracy of the first – and other – excited vibrational state is lifted (dynamic Jahn Teller effect). The component vibrational wave functions now transform according to the species A_1' and E', and the Raman transitions from the ground state to these two levels are both related to aymmetric tensors. This again as a result of the occurrence of species A_2' in the antisymmetric product. The occurrence of degenerate electronic ground states is very limited, and such a degeneracy can be lifted if a non-totally symmetric vibration interacts with the ground state, as is the case in the static Jahn Teller effect (or Renner Teller effect if the molecule is diatomic).

The occurrence of degenerate electronic states is more general among the sets of excited states, and in order to test the above, it is necessary to carry out Raman scattering experiments of excited states. Such states may be populated using optical pumping techniques, and this area of Raman spectroscopy has not yet been studied.

IV-6. Rotational Levels and Wave Functions

One of the more important properties of rotating bodies is the various moments of inertia. For a diatomic molecule with masses m_1 and m_2 separated by a distance r, the classical energy of rotation is given by $E_{\text{rot}} = \frac{1}{2} I\omega^2$, where I is the moment of inertia about the axis of rotation which is perpendicular to the internuclear axis and ω is the angular velocity. If the m_1 is at a distance of r_1 of the center of gravity, and m_2 at the distance r_2, then:

$$I = m_1 r_1^2 + m_2 r_2^2 = \mu r_2^2 \qquad \text{(IV, 6-1a)}$$

where the reduced mass

$$\mu = \frac{m_1 m_2}{m_1 + m_2}. \qquad \text{(IV, 6-1b)}$$

The angular momentum of this diatomic system is given by

$$P = I\omega \qquad \text{(IV, 6-2)}$$

and substitution of Equation (IV, 6-2) in the equation for the rotational energy gives:

$$E_{\text{rot}} = \frac{P^2}{2I}. \qquad \text{(IV, 6-3)}$$

For a rotating polyatomic molecule, there are three moments of inertia – I_x, I_y and I_z. It is always possible to select three mutual orthogonal axes about which the moments I_x, I_y and I_z have either a maximum or a minimum value. The moments associated with these axes are called the principal moments of inertia and they are denoted by

I_x^0, I_y^0 and I_z^0. In the zeroth order Born-Oppenheimer approximation, the total wave function is split into an electronic, a vibrational and a rotational part, and the Hamiltonian of the rigid rotator may then be written as:

$$\mathscr{H} = \tfrac{1}{2}\left(\frac{P_x^2}{I_x^0} + \frac{P_y^2}{I_y^0} + \frac{P_z^2}{I_z^0}\right). \tag{IV, 6-4}$$

Here P_x is the angular momentum about the x-axis. For the linear diatomic molecule $I_x^0 = I_y^0$ and $I_z^0 = 0$. (The z-axis is the line which connects the two mass points.) The wave equation in polar coordinates of the rotating system can be written as:

$$\frac{1}{r^2}\left[\frac{1}{\sin\theta}\frac{\partial}{\partial\theta}(\sin\theta)\frac{\partial\vartheta(J,M)}{\partial\theta} + \frac{1}{\sin^2\theta}\frac{\partial^2\vartheta(J,M)}{\partial\varphi^2}\right] + \frac{8\pi^2\mu}{h^2}E_{\text{rot}}\theta(J,M) = 0 \tag{IV, 6-5}$$

and the energy of the rotational states of the diatomic rigid rotator is found to be:

$$F_J = E_{\text{rot}} = BJ(J+1) \quad \text{where} \quad B = \frac{h^2}{8\pi^2 I_x^0}; \qquad I_x^0 = I_y^0 \tag{IV, 6-6}$$

and the rotational quantum number J can assume the values $J=0$, 1, 2, 3.... The magnitude of the angular momentum in terms of J is:

$$|\mathbf{P}| = \frac{h}{2\pi}\sqrt{J(J+1)}. \tag{IV, 6-7}$$

The wave functions $\vartheta(J, M)$ also follow from (IV, 6-5), they are:

$$\vartheta(J,M) = \frac{1}{\sqrt{2\pi}}\sqrt{\frac{(2J+1)(J-|m|)!}{2(J+|m|)!}}\, e^{im\varphi}P_J^{|m|}(\cos\theta). \tag{IV, 6-8}$$

Thus, for every value of $J>0$, there are several functions corresponding to different values of the magnetic quantum number M where $M=J$, J-1 .... J. There are thus $2J+1$ values of M, and $M(h/2\pi)$ is the component of the angular momentum in the direction of an electric or magnetic field applied parallel to the z-axis. The $P_J^{|m|}(\cos\theta)$ are normalized surface harmonics. Some simple examples are:

$$P_0^0 = 1; \qquad P_1^0 = \cos\theta,\ P_1^1 = \sqrt{1-\cos^2\theta}; \qquad P_2^0 = \tfrac{3}{2}\cos^2\theta - \tfrac{1}{2}.$$

For molecules having tetrahedral or higher symmetry, all three principal moments are equal. Thus, for spherical top molecules $I_x^0 = I_y^0 = I_z^0$. For asymmetric top molecules $I_x^0 \neq I_y^0 \neq I_z^0$, and a symmetric top is characterized by $I_x^0 = I_y^0 \neq I_z^0 \neq 0$. The direction of the principal moments coincides with the direction of rotational axis of high symmetry of the molecule. If the axis of highest symmetry is threefold or higher, then two of the moments of inertia are identical, and the molecule is a symmetric top. A less symmetrical molecule is an asymmetric top except if two of the moments are accidentally equal in value. The energy of the rotational states for symmetric top molecules is:

$$F_{J,K} = B[J(J+1) + (A/B - 1)K^2] \tag{IV, 6-9}$$

where

$$A = \frac{h^2}{8\pi^2 I_z^0} \quad \text{and} \quad B = \frac{h^2}{8\pi^2 I_x^0} \quad \text{with} \quad I_x^0 = I_y^0 .$$

K is a quantum number with the values:

$$K = J, J-1 \ldots -J$$

and $K(h/2\pi)$ is the component of the angular momentum in the z-direction. The energy of the rotational states of a spherical top is similar to that of the linear diatomic molecule so that:

$$F_{J,K} = BJ(J+1) \tag{IV, 6-10}$$

and the wave function is obtained by replacing the quantum number m by K in Equation (IV, 6-8). The rotational energy of asymmetric top molecules cannot be derived in a direct way, but use can be made of tabulated values of the functions $F_{J,\tau}(\mathbf{K})$ where the quantum number τ assumes the values:

$$\tau = J, J-1, \ldots - J$$

and the asymmetry parameter $\mathbf{K}$ is given by:

$$\mathbf{K} = \frac{2B - A - C}{A - C}$$

where

$$A = \frac{h^2}{8\pi^2 I_x^0}; \qquad B = \frac{h^2}{8\pi^2 I_y^0}; \qquad C = \frac{h^2}{8\pi^2 I_z^0}. \tag{IV, 6-11}$$

The energy levels of the rotational states for the various types of molecules are also given in Table IV-8.

TABLE IV-8
Energy of rotational levels for various molecules

Type	Moment of inertia	Energy	Degeneracy
linear	$I_x^0 = I_y^0 \neq 0, I_z^0 = 0$	$BJ(J+1)$	$2J+1$
Spherical top	$I_x^0 = I_y^0 = I_z^0$	$BJ(J+1)$	$(2J+1)^2$
Symmetric top	$I_x^0 = I_y^0 \neq I_z^0$	$B\left[J(J+1) + \left(\frac{A}{B} - 1\right)K^2\right]$	$2J+1\ (K=0)$ $2(2J+1)\ (\lvert K\rvert > 0)$
Asymmetric top	$I_x^0 \neq I_y^0 \neq I_z^0$ $I_x^0 > I_y^0 > I_z^0$	$\frac{A+C}{2}J(J+1) + \frac{A-C}{2}F_{J,\tau}(\mathbf{K})$	$2J+1$

IV-7. The Rotational Raman Effect

Rotational Raman transitions can occur from the electronic ground state of the molecule to the rotational states but it may also be possible that the transitions terminate on rotational levels of an excited vibrational state. There is also the process whereby electronic excited states are involved, but much of the experimental data are related to rotational effects on the electronic ground state.

If the electronic ground state is denoted by g, then the transition v_m, J, $M \leftrightarrow v_m$, J', M' represents a pure rotational one, and v_m, J, $M \leftrightarrow v'_m$ J' M' is of the vibrational rotational type. For an element of the tensor $\alpha_{P\Sigma}$ where P and Σ are laboratory axis X, Y and Z, the following expression can be written:

$$(\alpha_{P\Sigma}) = \langle \varphi_{v'_m}(\mathbf{Q}_m)\, \vartheta(J', M')|\, (\hat{\alpha}^{Q_m}_{\varrho\sigma})_{kk}\, |\varphi_{v_m}(\mathbf{Q}_m)\, \vartheta(J, M)\rangle . \qquad \text{(IV, 7-1)}$$

It is assumed that the wave function $\varphi_{v'_m}(\mathbf{Q}_m)$ belongs to a vibrational state of the electronic ground state g. The Cartesian axis ϱ and σ (x, y or z) are molecule fixed axis, and for the rotation molecule, the position of these axis changes continuously with respect to the laboratory fixed axis X, Y and Z. It is always possible to relate one set to the other by means of the direction cosines so that:

$$(\hat{\alpha}_{P\Sigma}) = \sum_{\varrho\sigma} \phi_{P\varrho}\phi_{\Sigma\sigma}\, (\hat{\alpha}^{Q_m}_{\varrho\sigma})_{kk} \qquad \text{(IV, 7-2)}$$

and substitution of the relation in Equation (IV, 7-1) leads to:

$$(\alpha_{P\Sigma}) = \langle \varphi_{v'_m}(\mathbf{Q}_m)|\, (\hat{\alpha}^{Q_m}_{\varrho\sigma})_{kk}\, |\varphi_{v_m}(\mathbf{Q}_m)\rangle\, \langle \vartheta(J', M') \sum_{\varrho\sigma} \phi_{P\varrho}\phi_{\Sigma\sigma}\vartheta(J, M)\rangle . \qquad \text{(IV, 7-3)}$$

It thus appears that the selection rules for the rotational transitions are related to the for non-vanishing of the second integral, assuming that the vibrational integral $\langle \varphi_{v'_m}(Q)|\, (\hat{\alpha}^{Q_m}_{\varrho\sigma})_{kk}\, |\varphi_{v_m}(Q)\rangle$ is different from zero. It is seen that for the pure rotational transitions, the wave functions $\varphi_{v'_m=0}(Q)$ and $\varphi_{v_m=0}(Q)$ transform according to the totally symmetric species of the point group of the molecule. At the same time, for a spherical top molecule, the sum of the products of direction cosines equals unity, and the selection rule is $\Delta J = 0$. At present it is not readily seen what the selection rules are for more complicated situations for which the above product cannot be directly summed. Values of the matrix element $\langle \vartheta(J', M') \Sigma_{\varrho\sigma}\, \phi_{P\varrho}\, \phi_{\Sigma\sigma}\, \vartheta(J, M)\rangle$ have been calculated by Casimir*, the results of which are given in Table IV-9. Presently, it is of importance to realize that the elements of the tensor are those of the vibrational Raman effect which, in turn, are related to the polarizability tensor. Thus the elements of this tensor may be called an internal property of the scattering molecule and depend on the vibrational quantum number and even on the electronic quantum number.

The Hamiltonian for such a rotation system has also been considered by Hougen and it was shown that basis functions can be derived which reflect the internal coor-

* Casimir, H. B. G.: 1929, *Z. Physik* **59**, 623.

TAE

Selection rules for rotational transiti

Rotor type	Vibrational species	Non-zero component of polarizability (cartesian)	irreduci
Linear	Totally symmetric	$\alpha_{xx} + \alpha_{yy}, \alpha_{zz}$	α_0^0, α_0^2
	Degenerate	α_{yz}, α_{zx}	$\alpha_2^2, \alpha_{-2}^2$
			$\alpha_1^2, \alpha_{-1}^2$
Symmetric top	Totally symmetric	$\alpha_{xx} + \alpha_{yy}, \alpha_{zz}$	α_0^0, α_0^2
	Non-totally symmetric[e], non-degenerate	$\alpha_{xx} - \alpha_{yy}, \alpha_{xy}$	$\alpha_2^2, \alpha_{-2}^2$
	Degenerate	$\alpha_{xx} - \alpha_{yy}, \alpha_{xy}$	$\alpha_2^2, \alpha_{-2}^2$
		α_{yz}, α_{zx}	$\alpha_1^2, \alpha_{-1}^2$
Spherical top	Totally symmetric	$\alpha_{xx} + \alpha_{yy} + \alpha_{zz}$	α_0^0
	Doubly degenerate	$\alpha_{xx} + \alpha_{yy} - 2\alpha_{zz}$,	α_0^2
		$\alpha_{xx} - \alpha_{yy}$	$\alpha_2^2 + \alpha_-$
	Triply degenerate	$\alpha_{xy}, \alpha_{yz}, \alpha_{xz}$	$\alpha_2^2 - \alpha_-$
			$\alpha_1^2, \alpha_{-1}^2$
Asymmetric Top[f]	Totally symmetric	$\alpha_{xx}, \alpha_{yy}, \alpha_{zz}$	α_0^0, α_0^2
	Non-totally symmetric	α_{xy}	$\alpha_2^2, \alpha_1^2,$
		α_{xz}	α_{-1}^2, α_-
		α_{yz}	
Slightly Asymmetric Top[g]	Totally symmetric	$\alpha_{xx} + \alpha_{yy}, \alpha_{zz}, \alpha_{xy}$	$\alpha_0^0, \alpha_0^2,$
	Non-totally symmetric non-degenerate	$\alpha_{xx} - \alpha_{yy}, \alpha_{xy}$[h]	$\alpha_2^2, \alpha_{-2}^2$
		α_{xz}, α_{yz}[i]	$\alpha_1^2, \alpha_{-1}^2$

[a] Placzek, G. and Teller, E.: 1933, *Z. Physik* **83**, 209, 839.
[b] Strong
[c] Weak
[d] For pure rational spectra $J' + J'' \geqslant 0$.
[e] Only in molecules with fourfold axes.

dinates of the molecule and the normal rotational functions. The total wave function can be written as a product:

$$\Psi = \psi_{k, v_m} \chi(J, K, M). \tag{IV, 7-4}$$

ψ_{k, v_m} is a function of the internal coordinates of the system; k is an electronic and v_m is a vibrational quantum number. The rotational functions $\chi(J, K, M)$ are defined as the eigenfunctions to the eigenvalue equation:

$$\hat{J}^2 \chi(J, K, M) = h^2 J(J+1) \chi(J, K, M). \tag{IV, 7-5}$$

For integer values of J, the operator $\hat{J}^2$ can be expressed in terms of the Eulerian angles α, β and γ given in Figure IV-6 and this may also be done with the $\chi(J, K, M)$

·9

er Placzek and Teller[a])

tate of polarization	Selection rules
olarized	$\Delta J = 0^{b}, \pm 2$ ($J = 0 \leftrightarrow J' = 0$ only α_0^0)
epolarized	$\Delta J = 0^{c}, \pm 1, \pm 2$
	always $+ \leftrightarrow +, - \leftrightarrow -, + \leftrightarrow -$ ($J + J' \geqslant 0$)
olarized	$\Delta J = 0^{b}, \pm 1, \pm 2 : \Delta K = 0.$ ($J = 0 \leftrightarrow J' = 0$ only α_0^0)
epolarized	$\Delta J = 0, \pm 1, \pm 2 : \Delta K = \pm 2$
epolarized	$\Delta J = 0, \pm 1, \pm 2 : \Delta K = \pm 1$ ($J + J' \geqslant 2$)[d]
ompletely polarized	$\Delta J = 0^{b}$
epolarized	$\Delta J = 0, \pm 1, \pm 2$
Depolarized	$\Delta J = 0, \pm 1, \pm 2$ ($J + J' \geqslant 2$)[d]
	$\Delta J = 0, \pm 1, \pm 2$ ($J + J' \geqslant 2$) ($J = 0 \leftrightarrow J' = 0$ only α_0^0)
Polarized	$+ + \leftrightarrow + +, + - \leftrightarrow + -$
	$- + \leftrightarrow - +, - - \leftrightarrow - -$
Depolarized	$+ + \leftrightarrow + -, - + \leftrightarrow - -$
Depolarized	$+ + \leftrightarrow - -, + - \leftrightarrow - +$
Depolarized	$+ + \leftrightarrow - +, + - \leftrightarrow - -$
Polarized	$\Delta J = 0, \pm 1, \pm 2 : \Delta K = 0, \pm 1, \pm 2$ ($J = 0 \leftrightarrow J' = 0$ only α_0^0)
Depolarized	$\Delta J = 0, \pm 1, \pm 2 : \Delta K = \pm 2$
Depolarized	$\Delta J = 0, \pm 1, \pm 2 : \Delta K = \pm 1$ ($J + J' \geqslant 2$)[i]

[f] See Herzberg, } ref. six and seven at end of chapter.
[g] See Herzberg, }

[h] Only in molecules of point group C_{2v}, D_2, V_h.

[i] Only in molecules of point group C_v, C_2, C_{2h}, C_{2v}, D_2, V_h.

functions. The result is that the eigenfunctions $\chi(J, K, M)$ can be interpreted as the $\vartheta(J, M)$ eigenfunctions of the simple rigid rotator discussed in Section IV-6. For J half-integer the operator $\hat{J}^2$ is no longer expressible in terms of these angles, but the function χ can still be expressed in terms of the Euler angles. If it is assumed that the state has a definite angular momentum J, then the projection of this angular momentum on a specific axis is equal to K. If another set of axis is adopted, then the rotational eigenfunctions will still have definite momentum associated with them, but the projection of this angular momentum on the new z-axis shall generally be different. Since these projections must be quantized, it is therefore clear that any function having a given total angular momentum J can be expressed as a linear combination of states with angular momentum J and fixed projections M of this momentum on the new z-axis.

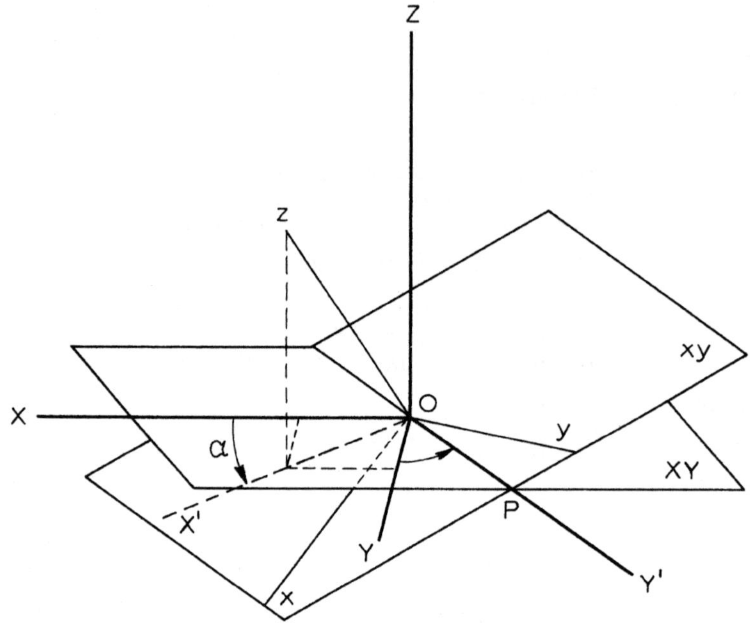

Fig. IV-6a. The Euler angles α, β and γ. Z, Y and X are space fixed axis and z, y, x are molecule fixed axis. The first rotation (α) around the Z-axis, carries OX over into OX' where OX' is the projection of Oz in the XY plane OP is the intersection of xy and XY plane and $OX' \perp OP$. Also $0 \leqslant \alpha \leqslant \pi$.

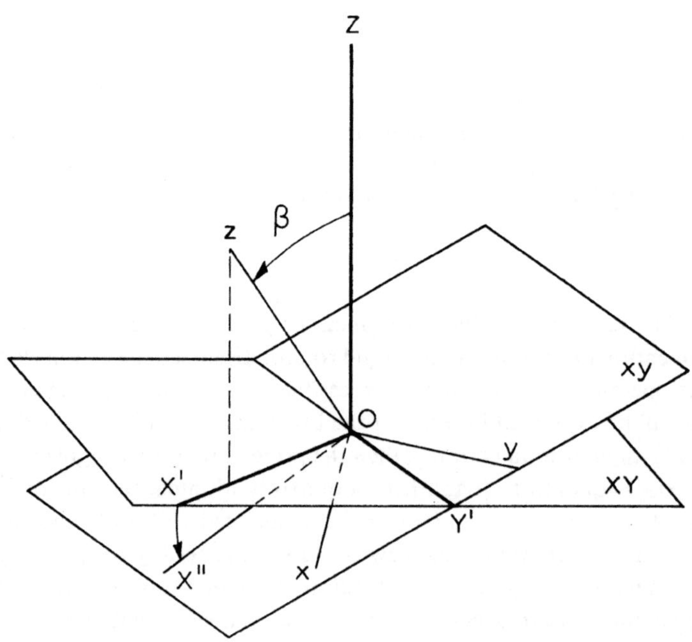

Fig. IV-6b. The second rotation (β) around the Y'-axis, carries OZ over into Oz and OX' into OX''. The line OX'' is now in the xy plane. ($0 \leqslant \beta \leqslant 2\pi$.)

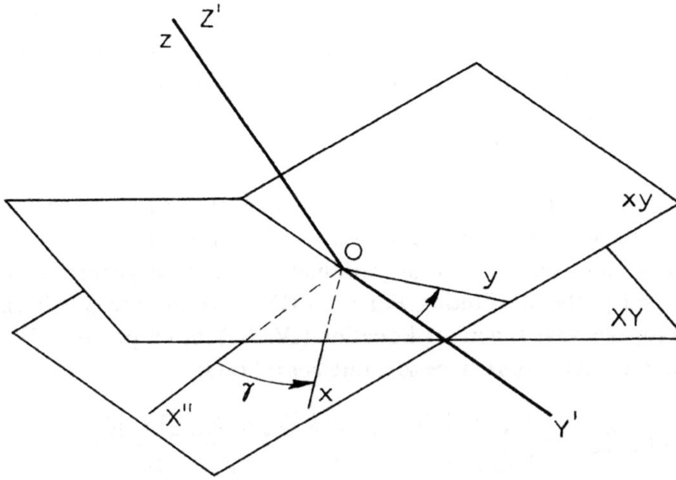

Fig. IV-6c. The third rotation γ around the $Z' = z$-axis, carries OX'' over into Ox and OY' into y. ($0 \leqslant \gamma \leqslant 2\pi$.)

If the former wave functions are denoted with $\chi(J, K)$, then:

$$\chi(J, K) = \sum_M D^J_{KM}(\alpha, \beta, \gamma)\, \chi'(J, M). \tag{IV, 7-6}$$

The transformation functions D^J_{KM} (α, β, γ) are functions of the three Euler angles relating the two sets of axis to each other. It may also be shown that these transformation functions will be eigenfunctions of the total angular momentum operator $\hat{J}$ having the fixed projections $K(h/2\pi)$ and $M(h/2\pi)$ along the initial and final z-axis respectively. The properties of these functions may be derived from the commutation rules for angular momentum components which, as it is well known, are independent, whether J is integer or half-integer. The transformation functions will be the correct eigenfunctions for the operator equation

$$\hat{J}^2 \chi(J, K, M) = \hbar^2 J(J+1)\, \chi(J, K, M) \tag{IV, 7-7}$$

and in the following, the notation D^J_{KM} shall be used rather than $\chi(J, K, M)$. The expression for an element $(\alpha_{\lambda\mu})_{kn}$ of the reducible spherical tensor (see (III, 5-4a) is:

$$(\alpha_{\lambda\mu})_{kn} = \frac{1}{h} \sum_r \frac{\langle\psi_k|\, r_\mu\, |\psi_r\rangle \langle\psi_r|\, r_\lambda\, |\psi_n\rangle}{\nu_{rk} - \nu_0} + \frac{[\lambda \leftrightarrow \mu]}{\nu_{rn} + \nu_0}. \tag{IV, 7-8}$$

In order to simplify Equation (IV 4, 7-8), it is necessary to separate those factors that are independent of the rotation of the molecule from those that depend on the rotations. The wave functions have already been split in this way, but the operator components are defined with respect to a space fixed coordinate system and thus depend on the internal as well as the rotational coordinates. This inconvenience can be removed in the following way. The operator components r_λ in the fixed coordinate system are then related to the operator components r_t of the rotating system by means of:

$$r_\lambda = \sum_t r_t D^1_{t\lambda}(\gamma, \beta, \alpha) \qquad \text{(IV, 7-9)}$$

where the transformation functions are discussed by Edmonds* and are identical to the rotational wave functions discussed earlier with the quantum number $J=1$. Now r_t depends only on the internal coordinates, and $D^1_{r\lambda}$ depends only on the Euler angles α, β and γ. By substituting Equation (IV, 7-9) and the previously discussed expressions for the wave functions Equation (IV, 7-6) into Equation (IV, 7-8), the rotational and internal factors can be separated as follows:

$$(\alpha_{\lambda\mu})^{nv'_mJ'}_{kv_mJ} = \frac{1}{h}\sum_r \sum_{t,t'} \left[\frac{\langle\psi_{k,v_m}| r_t |\psi_{r,v''_m}\rangle \langle\psi_{r,v''_m}| r_{t'} |\psi_{n,v'_m}\rangle}{\nu_{r,v''_m,J'';k,v_m,J} - \nu_0} \times \right.$$
$$\times \langle D^J_{KM}(k,v_m)| D^1_{t\lambda} |D^{J''}_{K''M''}(r,v''_m)\rangle \langle D^{J''}_{K''M''}(r,v''_m)| D^1_{t'\mu} |D^{J'}_{K'M'}(k,v'_m)\rangle +$$
$$\left. + \frac{[t \leftrightarrow t']}{\nu_{r,v''_m,J'';n,v'_m,J'} + \nu_0} \right]. \qquad \text{(IV, 7-10)}$$

The functions D_n etc. are the rotational wave functions with unspecified J, K and M.

In most cases that are experimentally interesting, the incident frequency ν_0 will be such that the change in the denominators of Equation (IV, 7-10) with the rotational quantum number be very small, and to a good approximation, this dependence can be neglected. Similar to the case of electronic and vibrational Raman effects, the closure theorem can be invoked because the rotational functions D^J_{KM} for all values of J, K and M constitute a complete orthonormal set of functions. The resulting expression is then:

$$(\alpha_{\lambda\mu})^{nv'_mJ'}_{kv_mJ} = \frac{1}{h}\sum_r \sum_{tt'} \left[\frac{\langle\psi_{k,v_m}| r_t |\psi_{r,v''_m}\rangle \langle\psi_{r,v''_m}| r_{t'} |\psi_{n,v'_m}\rangle}{\nu_{r,v''_m;k,v_m} - \nu_0} \times \right.$$
$$\left. \times \langle D^J_{KM}(k,v_m)| D^1_{t\lambda} D^1_{t'\mu} |D^{J'}_{K'M'}(n,v'_m)\rangle + \frac{[t \leftrightarrow t']}{\nu_{r,v''_m;n,v'_m} + \nu_0} \right]. \qquad \text{(IV, 7-11)}$$

The product $D^1_{t\lambda}\, D^1_{t'\mu}$ can now be expanded as:

$$D^1_{t\lambda} D^1_{t'\mu} = \sum_{TM'M} (2T+1) \begin{pmatrix} 1 & 1 & T' \\ t & t' & M' \end{pmatrix} D^{T'}_{M'M} \begin{pmatrix} 1 & 1 & T' \\ \lambda & \mu & M' \end{pmatrix}. \qquad \text{(IV, 7-12)}$$

The arguments of the functions in Equations (IV, 7-12) are different from those of the rotational wave functions. The two sets of arguments can be made identical by means

* Edmonds, A. R.: 1960, *Angular Momentum and Quantum Mechanics*, Princeton University Press, Princeton, N.J.

of the transformations

$$\begin{aligned} D^P_{t,Q}(\gamma, \beta, \alpha) &= e^{it\gamma}\, d^P_{tQ} e^{iQ\alpha} \\ &= e^{iQ\alpha}\, d^P_{Qt} e^{it\gamma} (-1)^{t+Q} \\ &= D^P_{Q,t}(\alpha, \beta, \gamma)\,(-1)^{t+Q}. \end{aligned} \tag{IV, 7-13}$$

The elements of the scattering tensor are $(\alpha_{\varrho\sigma})^{nv'_mJ'}_{kv_mJ}$ and in the discussion below, it is assumed that the rotational states are associated with vibrational states v_m and v'_m of the electronic state k.

$$(\alpha_{\varrho\sigma})^{nv'_mJ'}_{kv_mJ} = (\alpha_{\varrho\sigma})^{v'_mJ'M'K'}_{v_mJMK}. \tag{IV, 7-14}$$

Inserting Equation (IV, 7-12) in Equation (IV, 7-11) and by use of Equation (IV-7-13), the following result is obtained:

$$\begin{aligned} (\alpha_{\lambda\mu})^{v'_mJ'M'K'}_{v_mJMK} = \frac{1}{h}\sum_r\sum_{tt'}\Bigg[&\frac{\langle\psi_{k,v_m}|\, r_t\, |\psi_{r,v''_m}\rangle\langle\psi_{r,v''_m}|\, r_{t'}\, |\psi_{k,v'_m}\rangle}{\nu_{r,v''_m;k,v_m} - \nu_0} \times \\ &\times \sum_{T'N'N} \langle D^{J'v}_{K'M'}(v'_m)|\, D^{T'}_{NN'}\, |D^J_{KM}(v_m)\rangle \times (2T'+1) \times \\ &\times \begin{pmatrix} 1 & 1 & T' \\ t & t' & N' \end{pmatrix}\begin{pmatrix} 1 & 1 & T' \\ \lambda & \mu & N \end{pmatrix}(-1)^{N+N'} + \frac{[t' \leftrightarrow t]}{\nu_{r,v_m'';k,v_m} + \nu_0}\Bigg]. \end{aligned} \tag{IV, 7-15}$$

The product $|\, D^{T'}_{NN'}\, |D^J_{KM}\rangle$ may be expanded as:

$$|D^{T'}_{NN'}|\, D^J_{KM(v_m)}\rangle = \sum_{ULL'} (2U+1)\begin{pmatrix} T' & J & U \\ N & K & L \end{pmatrix} D^u_{LK}(v_m)\begin{pmatrix} T & J & U \\ N & M & L \end{pmatrix} \tag{IV, 7-16}$$

The orthonormality relation for the D^J_{KM} functions now requires that:

$$\langle D^{J_1}_{M'_1M_1}(v'_m)\;|\; D^{J_2}_{M'_2M_2}(v_m)\rangle = \frac{8\pi^2}{2J_1+1}\,\delta_{M'_1M'_2}\delta_{M_1M_2}\delta_{J_1J_2}. \tag{IV, 7-17}$$

By substituting Equation (IV, 7-16) and Equation (IV, 7-17) into Equation (IV, 7-15), the expression for the reducible spherical tensor becomes:

$$\begin{aligned} (\alpha_{\lambda\mu})^{J'M'K'}_{JMK} = \frac{1}{h}\sum_r\sum_{tt'}\Bigg[&\frac{\langle\psi_{k,v_m}|\, r_t\, |\psi_{r,v''_m}\rangle\langle\psi_{r,v''_m}|\, r_{t'}\, |\psi_{k,v'_m}\rangle}{\nu_{r,v''_m;k,v_m} - \nu_0} \times \\ &\times \sum_{T'N'N}\sum_{ULL'} (2T'+1)\begin{pmatrix} 1 & 1 & T' \\ t & t' & N' \end{pmatrix}\begin{pmatrix} 1 & 1 & T \\ \lambda & \mu & N \end{pmatrix}(-1)^{N+N'} \times \\ &\times (2U+1)\begin{pmatrix} T' & J & U \\ N & K & L' \end{pmatrix}\begin{pmatrix} T' & J & U \\ N' & M & L \end{pmatrix} \times \frac{8\pi^2}{2U+1} \times \\ &\times \delta_{L'K'}\delta_{LM'}\delta_{UJ'} + \frac{t \leftrightarrow t'}{\nu_{r,v''_m;k,v_m} + \nu_0}\Bigg] \end{aligned} \tag{IV, 7-18}$$

$$= \frac{1}{h} \sum_{r} \sum_{tt'} \left[\frac{\langle \psi_{k,v'm} | r_t | \psi_{r,v''m} \rangle \langle \psi_{r,v''m} | r_{t'} | \psi_{k,v'm} \rangle}{\nu_{r,v''m;k,v_m} - \nu_0} \times \right.$$

$$\times \sum_{T'N'N} (2T'+1) \begin{pmatrix} 1 & 1 & T' \\ t & t' & N' \end{pmatrix} \begin{pmatrix} 1 & 1 & T' \\ \lambda & \mu & N \end{pmatrix} (-1)^{N+N'} \times$$

$$\left. \times \begin{pmatrix} T' & J & J' \\ N & K & K' \end{pmatrix} \begin{pmatrix} T' & J & J' \\ N' & M & M' \end{pmatrix} \frac{8\pi^2}{2J'+1} + \frac{[t \leftrightarrow t']}{\nu_{r,v''m;k,v'm} + \nu_0} \right]. \tag{IV, 7-19}$$

It is now convenient to introduce the irreducible tensor components discussed in chapter III. Thus, according to Equation (III, 5-5):

$$\alpha_{\mathbf{Q}}^{\mathbf{K}} = \sum_{\lambda\mu} A_\lambda B_\mu C_{\lambda\mu\mathbf{KQ}} \,. \tag{IV, 7-20a}$$

According to Edmonds, this unitary transformation can also be written in terms of a $3j$ symbol:

$$\alpha_{\mathbf{Q}}^{\mathbf{K}} = \sum_{\lambda\mu} \alpha_{\lambda\mu} (-1)^{\mathbf{Q}} \begin{pmatrix} 1 & 1 & \mathbf{K} \\ \lambda & \mu & -\mathbf{Q} \end{pmatrix} \sqrt{2\mathbf{K}+1} \,. \tag{IV, 7-20b}$$

Substitution of the irreducible tensor components into the expression for the reducible spherical tensor leads to:

$$(\alpha_{\mathbf{Q}}^{\mathbf{K}})_{JMK}^{J'M'K'} = \frac{1}{h} \sum_{\lambda\mu} \sum_{r} \sum_{tt'} \sum_{T'N'N} \left[\frac{\langle \psi_{k,v'm} | r_t | \psi_{r,v''m} \rangle \langle \psi_{r,v''m} | r_{t'} | \psi_{k,v'm} \rangle}{\nu_{r,v''m;k,v_m} - \nu_0} \times \right.$$

$$\times (2T'+1) \begin{pmatrix} 1 & 1 & T' \\ t & t' & N' \end{pmatrix} \begin{pmatrix} 1 & 1 & T' \\ \lambda & \mu & N \end{pmatrix} (-1)^{N+N'} \times \begin{pmatrix} T' & J & J' \\ N & K & K' \end{pmatrix} \times$$

$$\left. \times \begin{pmatrix} T' & J & J' \\ N' & M & M' \end{pmatrix} \frac{8\pi^2}{2J'+1} \begin{pmatrix} 1 & 1 & \mathbf{K} \\ \lambda & \mu & -\mathbf{Q} \end{pmatrix} \sqrt{2\mathbf{K}+1} + \frac{[t \leftrightarrow t']}{\nu_{r,v''m;k,v_m} + \nu_0} \right]. \tag{IV, 7-21}$$

This expression can be simplified by carrying out the summation over λ and μ by introducing the orthogonality condition of the $3j$ symbols. The resulting expression is then:

$$(\alpha_{\mathbf{Q}}^{\mathbf{K}})_{JMK}^{J'M'K'} = \frac{1}{h} \sum_{N'tt'} (-1)^{\mathbf{Q}} \begin{pmatrix} 1 & 1 & \mathbf{K} \\ t & t' & N' \end{pmatrix} (-1)^{N'+\mathbf{Q}} \begin{pmatrix} \mathbf{K} & J & J' \\ -\mathbf{Q} & K & K' \end{pmatrix} \times$$

$$\times \begin{pmatrix} \mathbf{K} & J & J' \\ N' & M & M' \end{pmatrix} \frac{8\pi^2}{2J'+1} \sqrt{2\mathbf{K}+1} \sqrt{2J'+1} \times$$

$$\times \left[\frac{\langle \psi_{k,v_m} | r_t | \psi_{r,v''m} \rangle \langle \psi_{r,v''m} | r_{t'} | \psi_{k,v'm} \rangle}{\nu_{r,v''m;k,v_m} - \nu_0} + \frac{[t \leftrightarrow t']}{\nu_{r,v''m;k,v'm} + \nu_0} \right]. \tag{IV, 7-22}$$

A part of Equation (IV, 7-22) depends on the internal coordinates of the molecule, and it seems appropriate to relate that part to an internal irreducible tensor component

$(\alpha_{\mathbf{q}}^{\mathbf{K}})$ so that:

$$(\alpha_{\mathbf{q}}^{\mathbf{K}})_{\text{INT}} = \frac{1}{h} \sum_{tt'} (-1)^{\mathbf{q}} \begin{pmatrix} 1 & 1 & \mathbf{K} \\ t & t' & -\mathbf{q} \end{pmatrix} \sqrt{2\mathbf{K}+1} \times$$
$$\times \sum_{r} \left[\frac{\langle \psi_{k,v'm} | r_t | \psi_{r,v''m} \rangle \langle \psi_{r,v''m} | r_{t'} | \psi_{k,v'm} \rangle}{\nu_{r,v''m;k,vm} - \nu_0} + \frac{[t \leftrightarrow t']}{\nu_{r,v''m;k,v'm} + \nu_0} \right]$$
(IV, 7-23)

and the resulting expression for the irreducible tensor component $\alpha_{\mathbf{Q}}^{\mathbf{K}}$ related to the space fixed axis is then:

$$(\alpha_{\mathbf{Q}}^{\mathbf{K}})_{JMK}^{J'M'K'} = \sum_{\mathbf{q}'} 8\pi^2 \begin{pmatrix} \mathbf{K} & J & J' \\ -\mathbf{Q} & -K & K' \end{pmatrix} \begin{pmatrix} \mathbf{K} & J & J' \\ -\mathbf{Q}' & -M & M' \end{pmatrix} \times$$
$$\times (2J'+1)^{1/2} (\alpha_{\mathbf{q}}^{\mathbf{K}})_{\text{INT}}. \qquad \text{(IV, 7-24)}$$

This is the final expression for the tensor components. The selection rules for the rotational transitions are hidden in conditions for non-vanishing of the $3j$ symbols, assuming that the internal tensor components are different from zero. The $3j$ symbols are different from zero if $-\mathbf{Q}-K+K^1=0$ and also $-\mathbf{Q}-M+M^1=0$, while the triangular rule holds for the first rows. For Equation (IV, 7-24) this means that the sum of any of the two quantum numbers should be equal to or larger than the value of the third number. The quantities $\mathbf{K}$, M and M^1 refer to the projection of the total angular momentum on the space fixed z-axis, i.e. one of the laboratory axis for the Raman experiment and K^1 and K express projections along the molecular axis. The intensity of the rotational transitions is proportional to the square of the tensor elements, and in the calculation of the intensities, a summation has to be made over the M^1 substates. Also, the $2J+1$-fold degeneracy of the ground state should be taken into account. If Equation (IV, 7-24) is squared, then the following summations are important:

$$\sum_{\mathbf{q}} \sum_{\mathbf{q}'} \sum_{M'} \sum_{M''} \begin{pmatrix} \mathbf{K} & J & J' \\ -\mathbf{Q} & -K & K' \end{pmatrix}^2 \begin{pmatrix} \mathbf{K} & J & J' \\ -\mathbf{q} & -M & M' \end{pmatrix} \times$$
$$\times \begin{pmatrix} \mathbf{K} & J & J' \\ -\mathbf{q}' & -M & M'' \end{pmatrix} (\alpha_{\mathbf{q}}^{\mathbf{K}})_{\text{INT}}^2. \qquad \text{(IV, 7-25)}$$

The summation over the latter $3j$ symbols equals one, and for the intensity the following expression is obtained:

$$I_{\mathbf{Q}}^{\mathbf{K}} \approx (2J+1)(2J'+1) \begin{pmatrix} \mathbf{K} & J & J' \\ -\mathbf{Q} & -K & K' \end{pmatrix}^2 \sum_{\mathbf{q}} (\alpha_{\mathbf{q}}^{\mathbf{K}})_{\text{INT}}^2 N_J. \qquad \text{(IV, 7-26)}$$

This equation may be used to compute a rotational spectrum in terms of the $(\alpha_q^{\mathbf{K}})_{\text{INT}}$. A comparison of this equation and (IV, 7-3) shows that the selection rules for the transition are related to the conditions for non-vanishing of the $3j$ symbol $\begin{pmatrix} \mathbf{K} & J & J' \\ -\mathbf{Q} & -K & K \end{pmatrix}$.

For a spherical top molecule, the component α_0^0 transforms according to the totally symmetric species of the group T_d. Thus, $\mathbf{K}=0$ and $\mathbf{q}=0$. The former condition implies that $\Delta J=0$. If $J=0$ then $K=0$ and $\mathbf{Q}=0$. Thus, there does not exist rotational fine-structure of the Raman band which is normally assigned to the totally symmetric vibration of the molecule. At the same time, there does not exist rotational fine structure close to the Rayleigh line.

For a linear molecule in a non-degenerate electronic ground state, the pure rotational Raman spectrum is related to the spherical component α_0^0 only. The projection of the angular momentum on the internuclear axis is equal to zero and consequently the quantum number K and also K' equals zero. In the situation that the electronic ground state becomes degenerate, the component α_0^2 contributes also to the intensity of the rotational Raman lines. The occurrence of even antisymmetric tensors can become a reality if in the product of the representation of the ground state a species occurs to which the antisymmetric components belong. For a transition of the vibration rotation spectrum of a linear molecule these antisymmetric tensors cannot contribute. But for the case of the non-degenerate ground state it should be remembered that the component α_0^0 contributes only to the $J'=0 \leftarrow J=0$ transition and that the components α_0^0 and α_0^2 contributed to the other $J' \leftrightarrow J$ transitions ($\Delta J=0$). This has some interesting consequences with respect to the value of the depolarization ratio for a system of random oriented molecules in space (see the following section IV-8). For the transition $J'=0 \leftarrow J=0$ of the Q-branch of the vibration rotation spectrum of the linear molecule $\varrho_l=0$ but for all other transitions of this Q-branch $0<\varrho_l<\frac{3}{4}$.

For symmetrical top molecules, the components α_0^0 and α_0^2 transform according to the totally symmetric species of the appropriate point group. From the triangular rule of the $3j$ symbol, it follows that $\Delta J=0$ and $\Delta J=\pm 2$ respectively. The rotational states of the ground state of the molecule are at ambient temperatures sufficiently populated so that Raman transitions can occur from these excited states to higher lying states. If the initial state has a $J=1$ value, then with $\mathbf{K}=2$ the value of $J'=2$ and thus $\Delta J=1$. Thus for $J+J' \geqslant 2$, the series $\Delta J=\pm 1$ is also allowed (see also Table IV-9).

The rotational transitions are normally discussed in terms of branches. They are denoted by the letters O, P, Q, R, S and the corresponding ΔJ values are -2, -1, 0, $+1$, $+2$ respectively. The rotational structure of vibrational Raman bands of molecules is thus related to an O, P, Q, R and S branch, and the energy difference from either the center of the vibrational line or Rayleigh line is, according to Equation (IV, 6-9):

$$\begin{aligned}
&\text{For: } S \text{ branch} && F_J - F_{J+2} = -(4J+6)B \text{ cm}^{-1} \\
&\quad R \text{ branch} && F_J - F_{J+1} = -(2J+2)B \text{ cm}^{-1} \\
&\quad Q \text{ branch} && F_J - F_J \;\; = \;\; 0 \\
&\quad P \text{ branch} && F_J - F_{J-1} = \;\; 2JB \text{ cm}^{-1} \\
&\quad O \text{ branch} && F_J - F_{J-2} = \;\; (4J-2)B \text{ cm}^{-1}.
\end{aligned} \qquad \text{(IV, 7-27)}$$

It is seen that in the energy differences the quantum number K – given in Equations (IV, 6-9) – does not occur. The reason is that as a result of the non-vanishing of the components α_0^0 and α_0^2, an additional selection rule exists for the value for M' in the

$3j$ symbol. Thus $\Delta M=0$ and, consequently, there does not exist a dependence of the energy or energy difference on the quantum number K.

The Equation (IV, 7-26) for the intensity of the rotational transition is rather general and does not include the temperature dependence. First, using the tables of the $3j$ symbols, the rotational spectrum – for the possible values of **K** and **Q** – may be computed in terms of the parameters $(\alpha_{\mathbf{q}}^{\mathbf{K}})_{\mathrm{INT}}$. Thus, the ratio of intensities of the various branches is directly related to ratios of values of appropriate $3j$ symbols and the spectrum can thus be computed if the population of the excited states is taken into account. The elements of the polarizability tensor have been calculated for very simple molecules, and in relating the intensity of more complicated molecules to that of H_2, absolute values of the elements of the polarizability tensor of these molecules can be obtained. Secondly, the effect of electronic excited states can easily be considered. By introduction of electronic wave functions of excited states in Equations (IV, 7-14) to (IV, 7-24), the problem of finding the proper $(\alpha_{\mathbf{q}}^{\mathbf{K}})_{\mathrm{INT}}$ is reduced to the discussion given in Section IV-2 on the vibro-electronic Raman effect, and the effect of degenerate electronic states is that discussed in Section IV-5. The important result is that the rotational fine structure may be due to $(\alpha_{\mathbf{q}}^{\mathbf{K}})_{\mathrm{INT}}=\alpha_{1,0-1}^{1}$, and $\Delta J=\pm 1$ transitions can thus be related to anti-symmetric tensors. For molecules belonging to the point groups C_6, C_4, C_3, C_2, C_i and C_s, the intensity of the R and P branches of the Rayleigh line and of the fine structure of totally symmetric vibrations may be due to the occurrence of antisymmetric tensors. For all other point groups, the antisymmetric tensors can only contribute to the intensity of the rotation vibration spectrum of modes which are not totally symmetric. The possibility of observing such antisymmetric tensors is limited because of the large energy differences of excited states and ground state of most molecules.

IV-8. Depolarization Ratios

In the discussions of the various scattering processes, emphasis was placed on the activity of the Raman tensor. But only in solids (through a suitable choice of directions of major crystallographic axis and directions of propagation of incident and scattered light) can the individual elements of the tensor be measured. For gases and liquids, the orientation of molecules with respect to a laboratory fixed axis system is completely random, and the polarization properties of Raman bands are thus related to the averaged values of the Raman tensor – see chapter II-4. Raman spectra can be measured in different ways. First, there is the method that the direction of observation is perpendicular to the direction of propagation of the incident laser beam, and secondly, both these directions are parallel, and forward or backward scattering spectra are observed.

In Figure IV-7 a situation is depicted for the 90° scattering experiment, and it is assumed that the scattering atom, ion or molecule is located at the origin of the Cartesian coordinate system. Also, the axis of highest symmetry of the molecule is parallel to the z-axis. The incident beam is propagating along the y-axis and the elec-

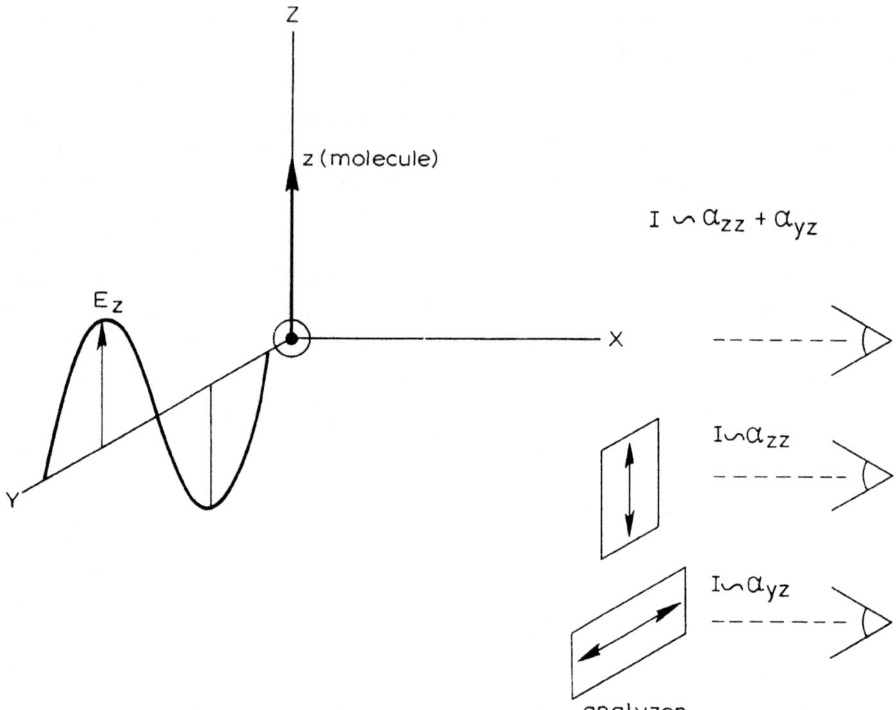

Fig. IV-7a. Measurement of tensor component of a single scattering molecule located at the origin.

tric vector can be parallel to either the z-axis or x-axis. The induced moments of interest are $\mathbf{M}_x$ and $\mathbf{M}_y$ if the scattered light is observed along the z-axis. However, if the Rayleigh and Raman radiation is observed along the x-axis, then the moments $\mathbf{M}_z$ and $\mathbf{M}_y$ contribute to the scattered light.* The total radiation emitted per unit solid angle in the x-direction is according to Equation (I, 5-24)

$$I = \frac{2\pi^3 (\nu - \nu_0)^4}{c^3} (\mathbf{M}_z^2 + \mathbf{M}_y^2) \qquad \text{(IV, 8-1)}$$

where

$$\mathbf{M}_y = \alpha_{yx}\mathbf{E}_x + \alpha_{yy}\mathbf{E}_y + \alpha_{yz}\mathbf{E}_z \qquad \text{(IV, 8-2a)}$$

$$\mathbf{M}_z = \alpha_{zx}\mathbf{E}_x + \alpha_{zy}\mathbf{E}_y + \alpha_{zz}\mathbf{E}_z\,. \qquad \text{(IV, 8-2b)}$$

For *circular-polarized* incident light propagating along the y-axis:

$$\left.\begin{aligned}\mathbf{M}_y &= \alpha_{yx}\mathbf{E}_x + \alpha_{yz}\mathbf{E}_z\\ \mathbf{M}_z &= \alpha_{zx}\mathbf{E}_x + \alpha_{zz}\mathbf{E}_z\end{aligned}\right\} \mathbf{E}_x = \mathbf{E}_z \neq 0\,, \qquad \mathbf{E}_y = 0 \qquad \text{(IV, 8-3)}$$

and for linear-polarized incident light:

$$\left.\begin{aligned}\mathbf{M}_y &= \alpha_{yz}\mathbf{E}_z\\ \mathbf{M}_z &= \alpha_{zz}\mathbf{E}_z\end{aligned}\right\} \mathbf{E}_z \neq 0; \qquad \mathbf{E}_x = \mathbf{E}_y = 0 \qquad \text{(IV, 8-4a)}$$

* See Figure I-2,4.

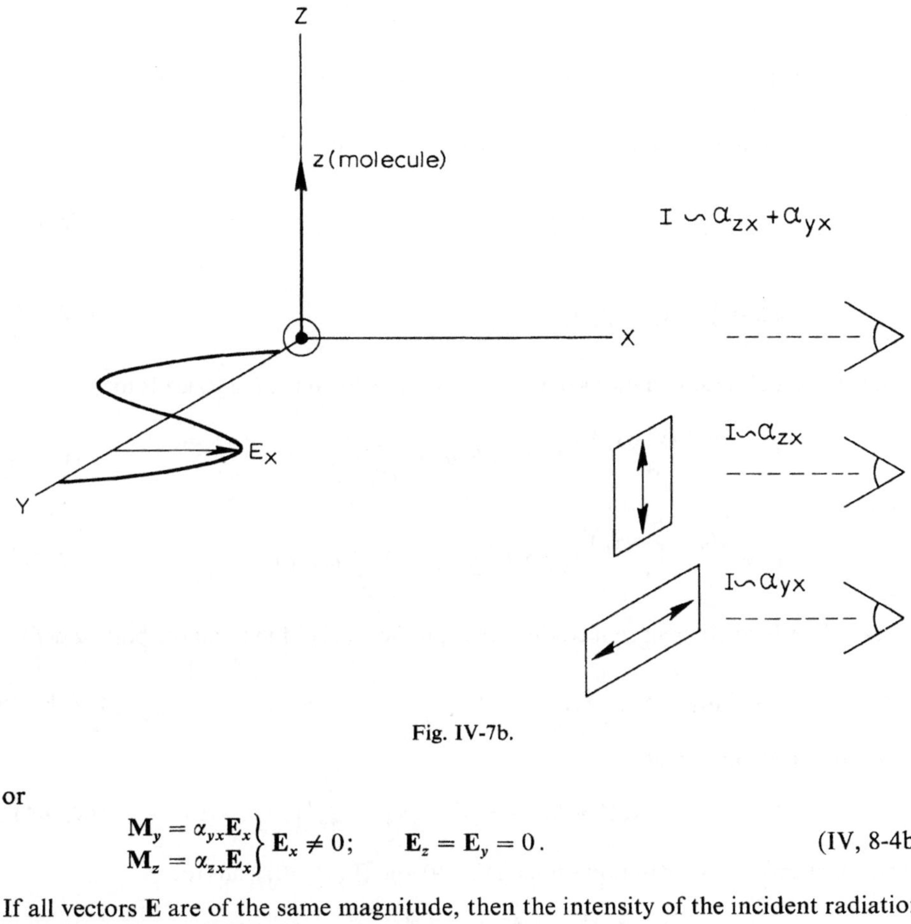

Fig. IV-7b.

or

$$\left.\begin{array}{l}\mathbf{M}_y = \alpha_{yx}\mathbf{E}_x\\ \mathbf{M}_z = \alpha_{zx}\mathbf{E}_x\end{array}\right\} \mathbf{E}_x \neq 0; \qquad \mathbf{E}_z = \mathbf{E}_y = 0\,. \tag{IV, 8-4b}$$

If all vectors **E** are of the same magnitude, then the intensity of the incident radiation is according to Equation (I, 1-7a):

$$I_0 = \frac{c}{8\pi}\,\mathbf{E}_\varrho^2 \quad (\text{where } \varrho = x,\ y \text{ or } z) \tag{IV, 8-5}$$

and the following are the total intensities emitted per unit solid angle in the direction of the x-axis if linear-polarized light is employed to excite the spectra:

$$I_z = \frac{16\pi^4(\nu-\nu_0)^4}{c^4}\,(\alpha_{yz}^2 + \alpha_{zz}^2)\,I_0 \qquad (\mathbf{E}_x = 0\,, \mathbf{E}_z \neq 0) \tag{IV, 8-6}$$

and

$$I_x = \frac{16\pi^4(\nu-\nu_0)^4}{c^4}\,(\alpha_{yx}^2 + \alpha_{zx}^2)\,I_0 \qquad (\mathbf{E}_x \neq 0\,, \mathbf{E}_z = 0)\,. \tag{IV, 8-7}$$

For N molecules oriented in such a way that their orientation in space is identical, the intensities per scattering center are additive, and the total intensity is obtained by multiplying the expressions by N. For gases and liquids, the molecules are not oriented in the same way, and the average of the tensor elements $\overline{\alpha_{nm}^2}$ (where n and m refer to the experimental axis X, Y or Z) over all orientations in space replaces $\alpha_{\varrho\sigma}^2$ in the expressions for the intensities.

The method to find these averages has been outlined in Chapter II-4.

Thus:

$$\overline{\alpha_{nm}^2} = \sum_i \tfrac{1}{15} \alpha_i^2 - \tfrac{1}{15} \sum_{i<j} \alpha_i \alpha_j \qquad \text{(IV, 8-8a)}$$

and

$$\overline{\alpha_{nn}^2} = \tfrac{1}{5} \sum \alpha_i^2 + \tfrac{2}{15} \sum_{i<j} \alpha_i \alpha_j \qquad \text{(IV, 8-8b)}$$

Substitution of these equations in Equations (IV, 8-6) and (IV, 8-7) leads to:

$$I_z = \frac{16\pi^4 (\nu - \nu_0)^4}{c^4} \tfrac{1}{15} N \left(4 \sum_i \alpha_i^2 + \sum_{i<j} \alpha_i \alpha_j\right) I_0 \qquad \text{(IV, 8-9)}$$

and

$$I_x = \frac{16\pi^4 (\nu - \nu_0)^4}{c^4} \tfrac{1}{15} N \left(2 \sum_i \alpha_i^2 - 2 \sum_{i<j} \alpha_i \alpha_j\right) I_0 . \qquad \text{(IV, 8-10)}$$

Presently it is convenient to introduce the spherical part of the polarizability α defined as:

$$\alpha = \tfrac{1}{3} (\alpha_1 + \alpha_2 + \alpha_3) \qquad \text{(IV, 8-11a)}$$

and the anisotropy β where

$$\beta = \tfrac{1}{2} [(\alpha_1 - \alpha_2)^2 + (\alpha_2 - \alpha_3)^2 + (\alpha_3 - \alpha_1)^2] . \qquad \text{(IV, 8-11b)}$$

Introduction of α and β in Equations (IV, 8-9) and (IV, 8-10) leads to:

$$I_z = \frac{16\pi^4 (\nu - \nu_0)^4}{c^4} N I_0 \frac{45\alpha^2 + 7\beta^2}{45} \qquad \text{(IV,8-12)}$$

$$I_x = \frac{16\pi^4 (\nu - \nu_0)^4}{c^4} N I_0 \frac{2\beta^2}{15} \qquad \text{(IV, 8-13)}$$

The depolarization ratio for Raman bands may now be defined as I_x/I_z (see Figure IV-8a).

Thus:

$$\varrho_n = \frac{6\beta^2}{45\alpha^2 + 7\beta^2} . \qquad \text{(IV, 8-14)}$$

On the other hand, it is also possible to analyse the scattered radiation. If an analyser is placed as indicated in Figure IV-8b, then, depending on the direction of preference for the electric vector of the analyser, the components $\overline{\alpha_{nn}^2}$ and $\overline{\alpha_{nm}^2}$ can be separated.

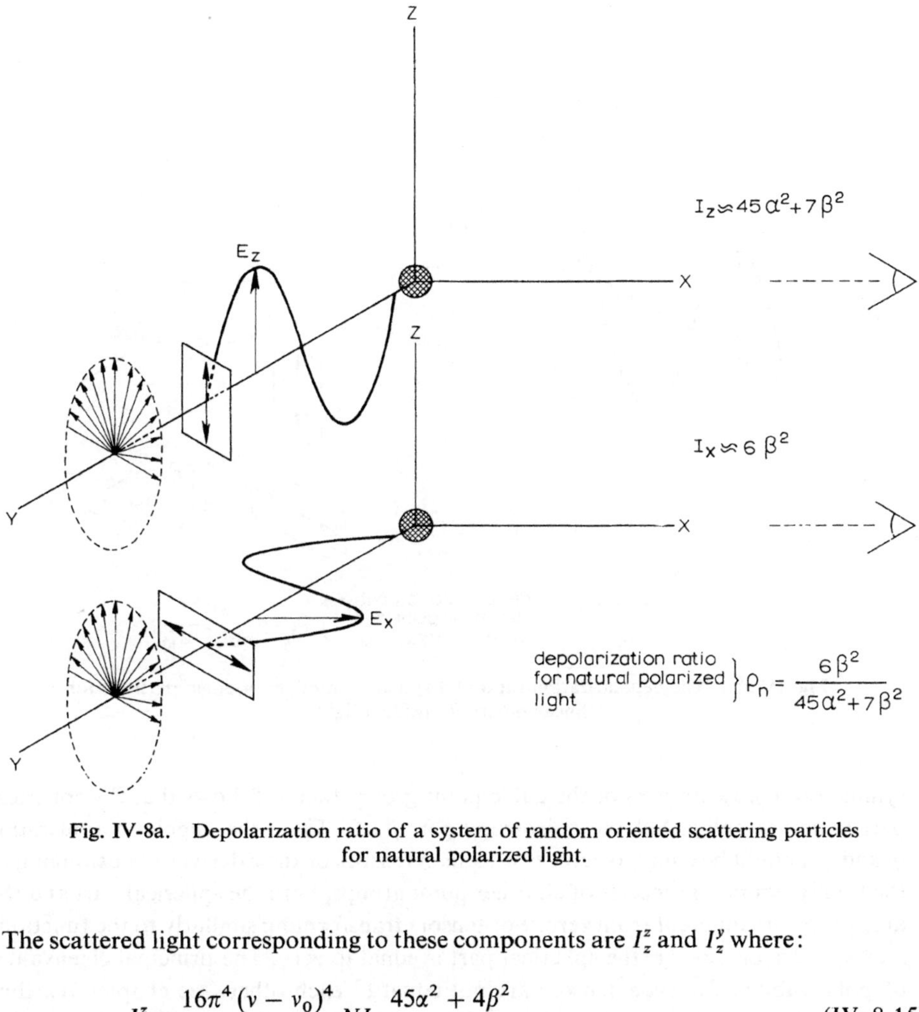

Fig. IV-8a. Depolarization ratio of a system of random oriented scattering particles for natural polarized light.

The scattered light corresponding to these components are I_z^z and I_z^y where:

$$I_z^z = \frac{16\pi^4 (\nu - \nu_0)^4}{c^4} N I_0 \frac{45\alpha^2 + 4\beta^2}{45} \tag{IV, 8-15}$$

and

$$I_z^y = \frac{16\pi^4 (\nu - \nu_0)^4}{c^4} N I_0 \frac{\beta^2}{15} \tag{IV, 8-16}$$

and the depolarization ratio can then be defined as:

$$\varrho_l = \frac{I_z^y}{I_z^z} = \frac{3\beta^2}{45\alpha^2 + 4\beta^2}. \tag{IV, 8-17}$$

For the Raman tensors of spectroscopic transitions, it is thus important to have at hand the values of α and β. If these tensors or their derivatives belong to the totally

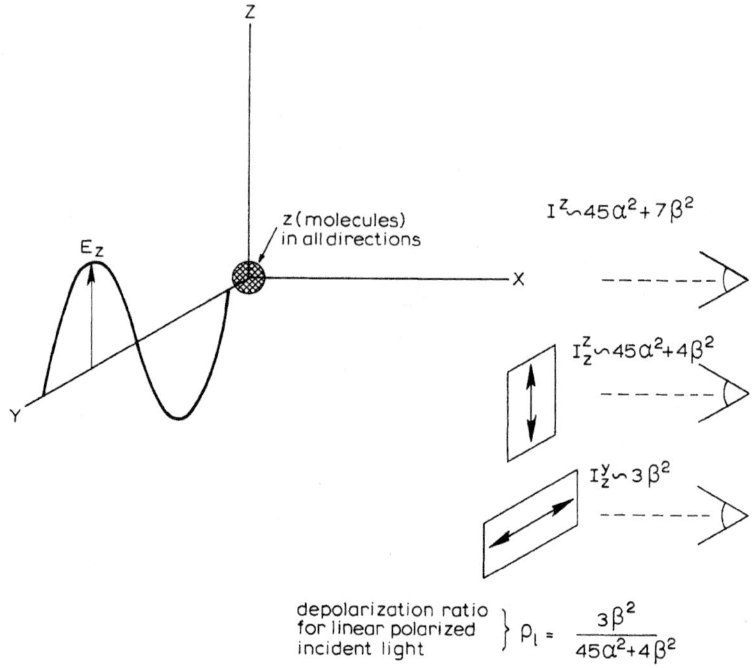

Fig. IV-8b. The depolarization ratio of a system in random oriented particles for linear polarized incident light.

symmetric representations of the cubic point group, then it follows that the spherical part is no-vanishing, but the anisotropy part $\beta=0$. Thus, the depolarization ratios ϱ_l and ϱ_n should be equal to zero. For Raman tensors or their derivatives belonging to the totally symmetric species of all other point groups, both the spherical part and the anisotropy are different from zero. For tensors transforming similarly to the functions $xx+yy-2zz$ or $xx-yy$, the spherical part is equal to zero. The principal eigenvalues of polarizability for such tensors are not equal to each other (see chapter II), thus $\beta \neq 0$. Such is also the case for a tensor $xy+yx$, and the important conclusions which can be drawn from the above that tensors or their derivatives associated with the non-totally symmetric species of the 32 point groups give rise to depolarization ratios $\varrho_l=\frac{3}{4}$ and $\varrho_n=\frac{6}{7}$. For the tensor belonging to the totally symmetric species $0<\varrho_l<\frac{3}{4}$ or $0<\varrho_n<\frac{6}{7}$, with the exception that $\varrho_n=\varrho_l=0$ if the point groups are the cubic groups. The above values for ϱ_l and ϱ_n are only valid if the tensors are symmetric. For a complete antisymmetric tensor (i.e. $\alpha_{\varrho\sigma}-\alpha_{\sigma\varrho}$), the intensities of corresponding transitions are calculated in a different way. From Equations (II, 4-15) and (II-16) it follows that the intensity I_z and I_x are given by:

$$I_z = \frac{16\pi^4(\nu-\nu_0)^4}{c^4} NI \tfrac{1}{10} |\alpha_{kl}|^2 \qquad \text{(IV, 8-18)}$$

and

$$I_x = \frac{16\pi^4 (\nu - \nu_0)^4}{c^4} NI \tfrac{1}{5} |\alpha_{kl}|^2 \qquad \text{(IV, 8-19)}$$

and the depolarization ratio becomes: $\varrho_n = 2$ and $\varrho_l = \infty$. For transitions which are related to only partly antisymmetric tensors, $\frac{6}{7} < \varrho_n < 2$ or $\frac{3}{4} < \varrho_l < \infty$. For the forward or backward scattering experiments the induced moments which can contribute to the intensity of the scattered radiation are $\mathbf{M}_z$ and $\mathbf{M}_x$, if incident and scattered light propagates along the y-axis and the vectors of the incident light are parallel to the experimental z- and x-axis. Thus:

$$\mathbf{M}_z = \alpha_{zx}\mathbf{E}_x + \alpha_{zz}\mathbf{E}_z \qquad \text{(IV, 8-20a)}$$

and

$$\mathbf{M}_x = \alpha_{xx}\mathbf{E}_x + \alpha_{xz}\mathbf{E}_z \,. \qquad \text{(IV, 8-20b)}$$

From the intensity ratio of I_z and I_x for a system of random oriented particles, it is not possible to obtain different values of the depolarization ratio for scattering tensors transforming according to the totally symmetric representation and those belonging to all other species. Such information becomes available if the scattered light is investigated with an analyser. The direction of preference of the electric vector can be chosen parallel to either the z-axis or the x-axis, and the intensity of the scattered light is thus:

$$I_z^z = \frac{16\pi^4 (\nu - \nu_0)^4}{c^4} \alpha_{zz}^2 I_0 \qquad (\mathbf{E}_z \neq 0, \mathbf{E}_x = 0) \qquad \text{(IV, 8-21a)}$$

and

$$I_z^x = \frac{16\pi^4 (\nu - \nu_0)^4}{c^4} \alpha_{xz}^2 I_0 \qquad (\mathbf{E}_z \neq 0, \mathbf{E}_x = 0) \,. \qquad \text{(IV, 8-21b)}$$

For N molecules oriented arbitrarily in physical space, the depolarization ratio is:

$$\varrho_l = \frac{I_z^x}{I_z^z} = \frac{3\beta^2}{45\alpha^2 + 4\beta^2} \,. \qquad \text{(IV, 8-22)}$$

In the derivations given above, the Cartesian tensor components were considered. In the expressions for the intensity of Roman lines of electronic and rotational scattering, such Cartesian components did not appear, but instead, use was made of the irreducible tensor components α_Q^K. The relation of one set to another is given by a unitary transformation (Table III-6). It is thus possible to state that:

$$|\alpha_{\varrho\sigma}|^2 = \Big(\sum_K \sum_q W_{\varrho\sigma}^{Kq} \alpha_q^K\Big) \Big(\sum_{K'} \sum_{q'} W_{\varrho\sigma}^{K'q'} \alpha_{q'}^{K'}\Big)^* \qquad \text{(IV, 8-23)}$$

where $W_{\varrho\sigma}^{Kq}$ is the unitary matrix linking the Cartesian basis set with the irreducible spherical basis set. The space fixed α_q^K's may be expressed in terms of α_Q^K which are defined in relation to a coordinate system fixed in the scattering system. If the Euler angles relating one system to the other are called α, β and γ, then, similar to Equation (IV, 7-6):

$$\alpha_q^K = \sum_Q \alpha_Q^K D_{Qq}^K (\alpha, \beta, \gamma) \,. \qquad \text{(IV, 8-24)}$$

Substitution of Equation (IV, 8-24) in Equation (IV, 8-23) gives:

$$|\alpha_{\varrho\sigma}|^2 = (\sum_K \sum_q \sum_Q W_{\varrho\sigma}^{Kq} \alpha_Q^K D_{Qq}^K (\alpha, \beta, \gamma)) (\sum_{K'} \sum_{Q'} \sum_{q'} W_{\varrho\sigma}^{K'q'} \alpha_{Q'}^{K'} D_{Q'q}^K (\alpha, \beta, \gamma)^*) \quad \text{(IV, 8-25a)}$$

$$= (\sum_K \sum_Q \sum_q W_{\varrho\sigma}^{Kq} \alpha_Q^K) (\sum_{K'} \sum_{Q'} \sum_{q'} W_{\varrho\sigma}^{K'q'} \alpha_{Q'}^{K'})^* \times$$

$$\times \sum_{KK'} \sum_{\sigma q'} D_{Qq}^K (\alpha, \beta, \gamma)\, D_{Q'q'}^{K'*} (\alpha, \beta, \gamma). \quad \text{(IV, 8-25b)}$$

The averaging of $\alpha_{\varrho\sigma}$ over all orientation in space is now reduced to an averaging over the angles α, β and γ of products of the type: $D_{Qq}^K(\alpha, \beta, \gamma) D_{Q'q'}^{K'}(\alpha, \beta, \gamma)$. The averaging is represented by:

$$\frac{\int\limits_{\alpha=0}^{2\pi} \int\limits_{\beta=0}^{\pi} \int\limits_{\gamma=0}^{2\pi} D_{\sigma q}^K (\alpha, \beta, \gamma)\, D_{Q'q'}^{K'*} (\alpha, \beta, \gamma)\, \mathrm{d}\alpha\, \mathrm{d} \sin\beta\, \mathrm{d}\gamma}{\int\limits_{\alpha=0}^{\pi} \int\limits_{\beta=0}^{2\pi} \int\limits_{\gamma=0}^{2\pi} \mathrm{d}\alpha \sin\beta\, \mathrm{d}\beta\, \mathrm{d}\gamma} \quad \text{(IV, 8-26a)}$$

The integrals have been discussed by Edmonds and

$$\overline{D_{Qq}^K (\alpha, \beta, \gamma)\, D_{Q'q'}^{K'*} (\alpha, \beta, \gamma)} = \delta(K, K')\, \delta(Q, Q')\, \delta(q, q'). \quad \text{(IV, 8-26b)}$$

Thus:

$$\overline{|\alpha_{\varrho\sigma}|^2} = \sum_K \sum_Q \sum_K |W_{\varrho\sigma}^{Kq}|^2 \frac{1}{2k+1} |\alpha_Q^K|^2. \quad \text{(IV, 8-27)}$$

If the constants $c_{\varrho\sigma}$ are introduced so that

$$C_{\varrho\sigma}^K = \sum_q |W_{\varrho\sigma}^{Kq}|^2 \quad \text{(IV, 8-28)}$$

then

$$\overline{|\alpha_{\varrho\sigma}|^2} = \sum_K \sum_Q C_{\varrho\sigma}^K \frac{1}{2K+1} |\alpha_Q^K|^2. \quad \text{(IV, 8-29)}$$

The coefficients $c_{\varrho\sigma}^K$ are directly obtained from Table III-6 and they are also given in Table IV-10. This table can also be used to calculate values of the depolarization ratios. If the antisymmetric tensors are the only tensors which contribute to the Raman

TABLE IV-10

Values of the coefficients $C_{\varrho\sigma}{}^K$

$C_{\varrho\sigma}{}^K$	K		
	0	1	2
$\varrho = \sigma$	$\frac{1}{3}$	0	$\frac{2}{3}$
$\varrho \neq \sigma$	0	$\frac{1}{2}$	$\frac{1}{2}$

intensity, then, according to Equation (IV, 8-6), Equation (IV, 8-7) and Equation (IV, 8-14):

$$\varrho_n = \frac{2\,\overline{(\alpha_{\varrho\sigma})^2}}{\overline{(\alpha_{\varrho\sigma})^2} + \overline{(\alpha_{\sigma\sigma})^2}} = 2\,. \qquad \text{(IV, 8-30)}$$

There is, of course, no difference between the derivations for ϱ_n and also ϱ_l using Cartesian or irreducible tensors, but the latter are more closely related to particular types of Raman scattering and the averaging procedure is less complicated.

Suggestions for Further Reading

Albrecht, A. C.: 1961, *J. Chem. Phys.* **34**, 1476.

Ballhausen, C. J.: 1962, *Introduction to Ligand Field Theory*, McGraw-Hill, New York.

Childs, M. S. and Longuet-Higgins, H. C.: 1961, *Phil. Trans. Roy. Soc.* **254 A**, 259.

Dieke, G. H.: 1968, *Spectra and Energy levels of Rare Earth Ions in Crystals*, Interscience Publishers, New York.

Elliott, R. J. and Loudon, R.: 1963, *Phys. Letters* **3**, 189.

Herzberg, G.: 1945, *Molecular Spectra and Structure I. Spectra of Diatomic Molecules*, D. Van Nostrand Comp., New York.

Herzberg, B.: 1950, *Molecular Spectra and Structure II. Infrared and Raman spectra of Polyatomic Molecules*, D. Van Nostrand Comp., New York.

Herzberg, G.: 1966, *Molecular Spectra and Structure III. Electronic Spectra of Polyatomic Molecules*, D. Van Nostrand Comp., New York.

Hougen, J. and Singh, S.: 1963, *Phys. Rev. Letters* **10**, 406.

Koningstein, J. A.: 1966, *J. Opt. Soc. Am.* **56**, 1405.

Landau, L. D. and Lifshitz, E. M.: 1958, *Quantum Mechanics*, Pergamon Press.

Murphy, W. F., Holzer, W. and Bernstein, H. J.: 1969, *Appl. Spectrosc.* **23**, 211.

Placzek, G.: 1934, *Marx Handbuch der Radiologie* (2d ed.), Vol. VI Academische Verlagsgesellschaft, Leipzig, p. 209.

Rasetti, F.: 1930, *Z. Physik* **66**, 646.

Savin, F. A. and Sobelman, J. J.: 1959, *Opt. i Spektroskopiya* **7**, 733.

Stoicheff, B. P.: 1959, in H. W. Thompson (ed.), 'High Resolution Raman Spectroscopy', *Advances in Spectroscopy* **1**, 91.

Wilson, E. B., Decius, J. C. and Cross, P. C.: 1955, *Molecular Vibrations*, McGraw-Hill, New York.

CHAPTER V

OTHER SCATTERING PROCESSES

The normal Raman effect, discussed in chapter IV is most important if Raman spectroscopy is applied as a tool to unravel structural aspects of atoms, molecules or ions. It was pointed out that the Raman process discussed there is a second order type process, and the intensity of the emission lines is weak. It is thus not so surprising that the introduction of the laser as an excitation source has revolutionized this particular field of spectroscopy.

The new excitation source has, however, certain properties which permitted a more detailed study of the interaction of light and matter. Thus, not only that much better signal to noise ratios were obtained for lines of the normal Raman effect, but also hyper, Stimulated and Inverse Raman spectra were induced and subsequently observed. The above three processes are in a way conneted with the type of lasers used in the experiments. Lasers may be divided into two categories: (1) giant pulse and (2) continuous wave (CW). A giant pulse laser delivers at least ~ 1 J in 10^{-8} s (with a peak power of 100 MW), while the power of a line of the CW argon ion laser is $\sim$1W. Improvements of the normal Raman spectra are mostly obtained with CW lasers, and the hyper, Stimulated and Inverse Raman spectra with giant pulse lasers. The explanation of this can be traced back to the strength of the electric fields associated with these unconventional light sources. Typical values are 10^{10} V cm^{-1} and this number may be compared to field strength of 10^{8-10} V cm^{-1} which binds the outer electrons in atoms, molecules or ions. The intense electric field of a focussed beam of such giant pulse laser can even cause the occurrence of ionization of air.

In the hyper Raman effect, two quanta of incident laser light give rise to a two photon absorption, and the Raman shifts occur at the energies $h(2\nu_0 \pm |\nu_{nm}|)$, and the Stimulated and Inverse Raman scattering are related, in a way, to the occurrence of induced absorption and emission. The two photon absorption process and the induced spectra can be achieved as a result of a strong perturbing electric field.

The scattering particle may also be exposed to excitation radiation of a frequency which is close to the frequency of an absorption band of the atom, molecule or ion. The Resonance-Raman effect is most conveniently studied with CW laser. Here care is taken that the concentration of the scattering entity does not reach a value where, even under moderate electric fields of $\sim 10^5$ V cm^{-1}, the atom, molecule or ion decomposes. This chapter closes with a discussion of the properties of this scattering process.

V-1. The Hyper Raman Effect

In the hyper Raman effect, emission of radiation occurs at the energy $2h\nu_0 \pm h\nu_{nk}$.* The process is depicted in the figures given in chapter I. In the discussion given in chapter I, it was already pointed out that Equation (I, 4-23f) is not complete in describing the HRE because the wave functions used to evaluate the induced moment M_{kn} were only corrected to first order. In order to carry out the full calculation, it is necessary to consider the matrix element:

$$\mathbf{M}_{kn} = \int \Psi_n^* M \Psi_k \, d\tau = \int (\Psi_n^0 + \Psi_k' + \Psi_k'')^* \, M \, (\Psi_k^0 + \Psi_k' + \Psi_k'') \, d\tau$$
$$= \mathbf{M}_{kn}^{(0)} + \mathbf{M}_{kn}^{(1)} + \mathbf{M}_{kn}^{(2)} \qquad \text{(V,1-1)}$$

where the wave function ψ_k'' is a second order correction to the wave function ψ_k. The expressions for the hyper Raman effect are associated with $M_{kn}^{(2)}$ so that:

$$\mathbf{M}_{kn}^{(2)} = \int \Psi_n'^* M \Psi_k \, d\tau + \int \Psi_n''^* M \Psi_k^0 \, d\tau + \int \Psi_k^{0*} M \Psi_n'' \, d\tau \,. \qquad \text{(V, 1-2)}$$

The integrals containing the wave functions ψ_k'' and ψ_n'' can only be evaluated if a computation is made of these functions. This was done in 1970 by Long and Satton, and the contribution of the hyper Raman scattering to the induced moment $M_{kn}^{(2)}$ is given by:

$$\mathbf{M}_{kn}^{(2)} = \frac{1}{h^2} \sum_{r,p} \left[\frac{(\mathbf{A}\cdot\mathbf{M}_{kp})\,\mathbf{M}_{pr}\,(\mathbf{A}\cdot\mathbf{M}_{rn})}{(\nu_{rn} - \nu_0)\,(\nu_{pk} + \nu_0)} + \frac{(\mathbf{A}\cdot\mathbf{M}_{kp})\,\mathbf{M}_{rn}\,(\mathbf{A}\cdot\mathbf{M}_{pr})}{(\nu_{rk} + 2\nu_0)\,(\nu_{pk} + \nu_0)} + \right.$$
$$\left. + \frac{(\mathbf{A}\cdot\mathbf{M}_{rp})\,\mathbf{M}_{rk}\,(\mathbf{A}\cdot\mathbf{M}_{pn})}{(\nu_{rn} - 2\nu_0)\,(\nu_{pn} - \nu_0)} \right] e^{-2\pi i(\nu_{kn} + 2\nu_0)t} +$$
$$+ \frac{1}{h^2} \sum_{rp} \left[\frac{(\mathbf{A}\cdot\mathbf{M}_{kp})\,(\mathbf{M}_{pr})\,(\mathbf{A}\cdot\mathbf{M}_{rn})}{(\nu_{rn} + \nu_0)\,(\nu_{pk} - \nu_0)} + \frac{(\mathbf{A}\cdot\mathbf{M}_{kp})\,(\mathbf{M}_{rn})\,(\mathbf{A}\cdot\mathbf{M}_{pr})}{(\nu_{rk} - 2\nu_0)\,(\nu_{pk} - \nu_0)} + \right.$$
$$\left. + \frac{(\mathbf{A}\cdot\mathbf{M}_{pr})\,(\mathbf{M}_{kp})\,(\mathbf{A}\cdot\mathbf{M}_{rn})}{(\nu_{rn} + 2\nu_0)\,(\nu_{pn} + \nu_0)} \right] e^{-2\pi i(\nu_{kn} - 2\nu_0)t} \,. \qquad \text{(V, 1-3)}$$

The second term of Equation (V, 1-3) leads to scattering if $\nu_{kn} - 2\nu_0 > 0$ (Klein's principle). Thus, $\nu_{kn} > 2\nu_0$. The Raman transition takes place between the states k and n, and it is seen that the frequency of the transition is greater than that of the incident radiation. Since ν_0 must necessarily be positive, it follows that ν_{kn} must also be positive, and this implies that the initial state k must be lying above the state n. The probability of observation of such an hyper Raman effect is rather limited because the excited state has to be sufficiently populated. The radiation of the moment associated with the first member of Equation (V, 1-3) is given by the relations: $\nu_{kn} + 2\nu_0 > 0$ or

* It is assumed here that the incident photons have the same energy.

$2\nu_0 > \nu_{kn}$. If ν_{kn} is negative, then this corresponds to hyper Raman Stokes emission at $2\nu_0 - |\nu_{kn}|$, and the hyper Raman anti-Stokes emission occurs at $2\nu_0 + |\nu_{kn}|$.

For an element of the scattering tensor, the following expression is obtained (see also chapter I) for the hyper Raman effect:

$$(\beta_{\varrho\sigma\sigma}) = \frac{1}{h^2}\sum_{rp}\left[\frac{(M_\sigma)_{kp}(M_\sigma)_{pr}(M_\varrho)_{rn}}{(\nu_{pk}+\nu_0)(\nu_{rn}-\nu_0)} + \frac{(M_\sigma)_{kp}(M_\varrho)_{pr}(M_\sigma)_{rn}}{(\nu_{rk}+2\nu_0)(\nu_{pk}+\nu_0)} + \right.$$

$$\left. + \frac{(\mathbf{M}_\sigma)_{kr}(\mathbf{M}_{\sigma'})_{rp}(\mathbf{M}_\varrho)_{pn}}{(\nu_{kr}-2\nu_0)(\nu_{pn}-\nu_0)}\right] \qquad \text{(V, 1-4)}$$

and this tensor is, in principle, an asymmetric tensor. Following the theory for the normal Raman effect, it can be demonstrated that if the states k and n are pure electronic states, all 27 elements of the scattering tensor may individually contribute. If only vibrational levels on the electronic ground state are involved in the hyper Raman effect, then it is convenient to introduce the hyper polarizability β. Analogous to the polarizability of the ground state α_{kk}, there may also be defined the hyper polarizability of the ground state β_{kk}, and this tensor is a symmetric tensor. For the vibrational hyper Raman effect, the hyper-polarizability may be expanded in a Taylor series of a particular normal coordinate Q_a. Thus:

$$\beta_{kk} = \beta^0 + \left(\frac{\partial\beta_{kk}}{\partial Q_m}\right)_{Q_m=0} + \frac{1}{2}\left(\frac{\delta^2\beta_{kk}}{\partial Q_m^2}\right)_{Q_m=0} Q_m + \ldots \qquad \text{(V, 1-5)}$$

The selection rules again show that $v'_m = v_m \pm 1$, and the Raman transition is only allowed if one of the elements of the tensor – or a combination thereof – belongs to a species of the point group to which the normal coordinate $\mathbf{Q}_m$ also belongs. As a direct result of the fact that the tensor has 10 elements, compared to six for the polarizability tensor, situations do arise where $\beta_{\varrho\sigma\sigma'}$ and a normal coordinate transform identically, but not $\alpha_{\varrho\sigma}$ and the said normal coordinate (see the character tables of the point groups). This is the main difference between the normal and hyper-vibrational Raman effect. The extension of the normal selection rules is important in that a complete picture is obtained of the vibrational frequencies of a molecule. If proper units are used, then typical values of the polarizability α is 10^{-24} cm^3, and β is 10^{-30} esc^{-1} cm^5 The induced dipole moment of the ground state of a molecule $\mathbf{M}$ can be formally expressed in the polarizability and hyper-polarizability. Thus:

$$\mathbf{M} = \alpha\mathbf{E}_\sigma + \tfrac{1}{2}\beta\,\mathbf{E}\,\mathbf{E} + \ldots. \qquad \text{(V, 1-6)}$$

It follows that electric fields of considerable strength have to be present in order that the first hyper-polarizability can contribute to the moment of the ground state. The electric fields of CW lasers are in the order of 10^5 V cm$^{-1} \approx 300$ esu and the contribution of β can thus be neglected. For Q-spoiled lasers the strength of the E-field reaches values of at least 10^7 V cm$^{-1} \approx 3 \times 10^4$ esu and thus, the contribution of β is of considerable importance. It is for this reason that the transformation properties of β

are also given in the character tables of the point groups. Presently, only a few vibrational hyper Raman spectra have been recorded. In general, a Q-spoiled ruby laser ($\lambda_0 = 6943$ Å) is used as the excitation source, and Raman shifts are measured from the frequency $2\nu_0 = 28\,806$ cm^{-1}. The emitted radiation from the Raman cell is focussed on the slit of a monochromator and the dispersed light is detected with a photomultiplier. The monochromator is set at fixed frequencies, and the number of photoelectrons from the photocathode is counted with a photon-counting system. The result of a large number (~25) of laser pulses per 50 cm^{-1} band pass is about sufficient to produce a reasonable spectrum of water as shown in Figure V-1.

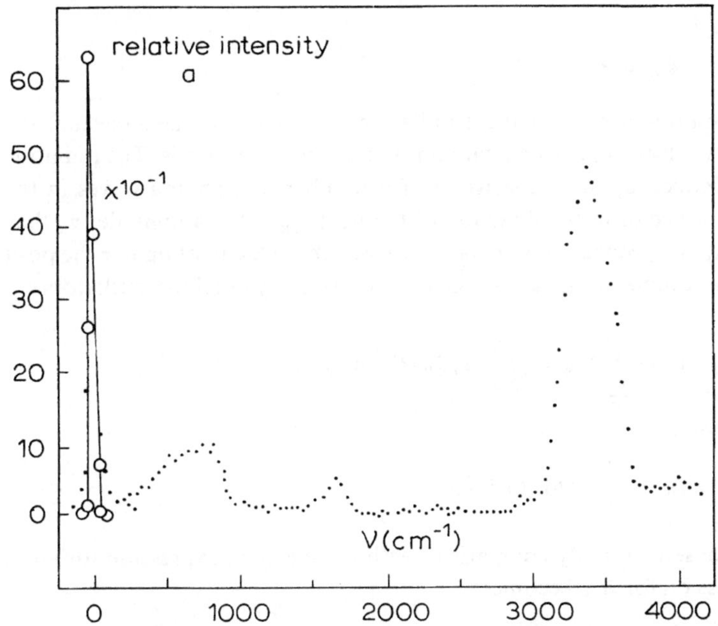

Fig. V-1. The hyper Raman spectrum of H_2O. (after Terhune, Maker and Savage)

Experimental difficulties are still present, and the technique of hyper Raman spectroscopy is not yet fully developed. Additional experimental results are necessary to make a judgement on the usefulness of these spectra.

The expansion of Equation (V, 1-5) can also be carried out to include the simultaneous interaction of three incident photons, but the probability of the process is low, and the scattering tensor is rather complicated.

V-2. The Stimulated Raman Effect

It was pointed out in chapter I that radiative processes can be either spontaneous or induced (stimulated), and in discussing the latter, the density of the radiation field is of importance. In all expressions for Raman scattering which originate from perturbation thereby where the molecule and the electric field are treated separately, $\varrho(\nu_0)$ or $\varrho(\nu_0+\Delta\nu)$ does not occur. Placzek, who used Dirac's treatment of the radiation field, derived expressions for the Raman effect which yielded the densities, and consequently, aspects of spontaneous and stimulated effects were known about thirty years before the observation of the stimulated Raman effect. In the quantum mechanical treatment of the radiation field, the molecule (or ion, atom) and the radiation field due to the incident photons of energy $h\nu_0$ are taken as one single system. Thus, the Hamiltonian is made up of contributions of the energy of the molecule $\mathscr{H}_M$ and the energy of the incident radiation $\mathscr{H}_R$. The Hamiltonian for the unperturbed system is given by:

$$\mathscr{H}_0 = \mathscr{H}_M + \mathscr{H}_R . \qquad \text{(V, 2-1)}$$

The assumption is made that the radiation field (and also the molecule) are contained in a cavity which is provided with antireflection coated walls. The radiation field may then be looked upon as consisting of a number of standing waves in the cavity. If $\mathbf{e}_R$ is a unit vector in the direction of the field, $\mathbf{q}_R$ is the amplitude, ν_R the frequency, $\mathbf{k}_R = 2\pi/\lambda_R$ the propagation or wave vector, c the velocity of light, $\mathbf{r}$ the position vector and V the volume of the cavity; then the total energy of all the oscillators is:

$$W = \sum_R W_R = \sqrt{\frac{8\pi}{V}}\, e_R \dot{q}_R(t) \sin\gamma_R$$

where

$$\gamma_R = \frac{2\pi\nu_R}{c}(\mathbf{k}_R \cdot \mathbf{r}) + \delta_R . \qquad \text{(V, 2-2)}$$

δ_R is a phase factor. By comparison with the classical expression for energy density, the expression for $\mathscr{H}_R$ becomes:

$$\mathscr{H}_R = \tfrac{1}{2} \sum_R (\dot{q}^2 + 4\pi\nu_R^2 q_R) \qquad \text{(V, 2-3)}$$

where the mass of all oscillators is taken as equal unity. The Hamiltonian for the unperturbed system is thus:

$$\mathscr{H} = \mathscr{H}_M + \tfrac{1}{2} \sum_R (\dot{q}^2 + 4\pi\nu_R^2 q_R^2) . \qquad \text{(V, 2-4)}$$

If the Schrödinger equation is solved, then the total energy is the sum of the energy of the molecule and the oscillators, so that:

$$W_{kR} = W_k + \sum_R (k_R + \tfrac{1}{2})\, h\nu_R . \qquad \text{(V, 2-5)}$$

k stands for the quantum numbers of the level of the molecule, and the oscillators are taken as simple harmonic type oscillators. The wave functions ψ_{kR} are product functions of the type:

$$\psi_{kR} = \psi_k \prod^{R} \varphi_R \tag{V, 2-6}$$

and also, it follows from $(\mathscr{H}_0 - W)\Psi = 0$ that:

$$\Psi = \sum_{kR} a_{kR}(t)\, \psi_{kR}\, e^{-iW_{kR}t}. \tag{V, 2-7}$$

$a_{kR}(t)$ is a time-dependent coefficient and $a^*_{kR}\, a_{kR}$ represent the probability of finding the system in the state kR at time t. At $t=0$, the molecule can be in the state k and the radiation field in state R_0. Thus, the radiation field of the incident light is then determined by the number of light quanta that occupy the individual vibrational eigenstates (N_1, N_2, $N_3 \ldots N_j$) of the harmonic oscillators. If an interaction between the molecule and the radiation field does not exist, the system remains in its given state; otherwise, the molecule may jump from quantum state k to n. The interaction energy is the same as given in the discussion of chapter I, thus $H^1 = -(\mathbf{E}\cdot\mathbf{M})$. First and second order perturbation theory may now be applied to calculate the a coefficients. Of particular interest are the results of the second order perturbation calculation:

$$a^{(2)}_{nR} = \sum_{rR'} \frac{(\mathscr{H}')^{kR_0}_{rR}\, (\mathscr{H}')^{rR'}_{nR}}{2(\nu_{rk} + \nu_{RR_0})} \left[\frac{1 - e^{2\pi i(\nu_{rn} + \nu_{RR'})t}}{2(\nu_{rn} + \nu_{RR_0})} - \frac{1 - e^{2\pi i(\nu_{nk} + \nu_{RR0})t}}{2(\nu_{nk} + \nu_{RR_0})} \right] \tag{V, 2-8}$$

where

$$(\mathscr{H}')^{kR_0}_{rR'} = \int \psi_{rR'} \mathscr{H}' \psi_{kR_0}\, \mathrm{d}\tau .$$

After the interaction has taken place, the radiation field is in state R and the molecule is in the quantum state n. R^1 and r are other states of the system and the summation goes over all these states. An inspection of Equation (V, 2-8) reveals that two of the cavity resonators have to change by $+1$ or -1 quantum steps. This is because the perturbation is linear in the coordinate q and the selection tule for harmonic oscillators is $\Delta k_R = \pm 1$. If the two oscillators are designated by α and β respectively, then the changes are represented as follows:

(a) $N'_\alpha = N_\alpha + 1$; $N'_\beta = N_\alpha + 1$. Thus, the occupation number of the αth and βth oscillator is increased by one, and this results in the presence of radiation of frequency ν_α and ν_β (double emission).

(b) $N'_\alpha = N_{\alpha-1}$; $N'_\beta = N_{\beta-1}$. The occupation number of light quanta is decreased by one, and here, a double absorption process takes place.

(c) $N'_\alpha = N_{\alpha-1}$; $N'_\beta = N_{\beta+}1$. A photon of energy $h\nu_\alpha$ is 'absorbed' (destructed) and a photon of energy $h\nu_\beta$ is 'emitted' (created).

(d) $N'_\alpha = N_{\alpha+1}$; $N'_\beta = N_{\beta-1}$. The same as under (c), but a photon of energy $h\nu_\alpha$ is created and $h\nu_\beta$ is destructed. The latter two processes represent a (Raman) scattering effect where (c) an incident photon of energy $h\nu_\alpha$ interacts with the molecule and is destructed, and a photon of frequency $h\nu_\beta$ is emitted.

Having made the assumption that the cavity vibrators carry out simple harmonic oscillations, the following matrix elements can be computed:

$$q_{K_R+1,K_R} = \int \varphi^*_{K_R} q \varphi_{K_R+1} \, \mathrm{d}\tau = \sqrt{\frac{(K_R+1)h}{8\pi^2 \nu_R}} \, e^{2\pi i \nu_R t}$$

and

$$\dot{q}_{K_R+1,K_R} = i \sqrt{\frac{(K_R+1)\, h\nu_R}{2}} \, e^{2\pi i \nu_R t} \tag{V, 2-9}$$

also

$$\dot{q}_{K_R-1,K_R} = \sqrt{\frac{K_R h}{8\pi^2 \nu_R}} \, e^{-2\pi i \nu_R t}$$

and

$$\dot{q}_{K_R-1,K_R} = i \sqrt{\frac{K_R h\nu_R}{2}} \, e^{-2\pi i \nu_R t} . \tag{V, 2-10}$$

A combination of Equations (V, 2-9), (V, 2-10) and (V, 2-2) with the expression for H^1 yields the following expression for the probability of finding the molecule in state n and the radiation field in state R, which is defined by the occupation numbers $(N_1, N_2, N_3 \ldots N_\alpha - 1 \ldots N_\beta + 1 \ldots N_j)$:

$$|a^{(2)}_{nR}|^2 = \frac{64\pi^2}{V^2} N_\alpha (N_\beta + 1)\, \nu_\alpha \cdot \nu_\beta \sin^2 \Gamma_\alpha \sin^2 \Gamma_\beta |M^{\alpha\beta}_{kn}|^2 \frac{\sin^2 \pi (\nu_{nk} + \nu_\alpha - \nu_\beta)\, t}{(\nu_{nk} + \nu_\alpha - \nu_\beta)^2}$$

where

$$M^{\alpha\beta}_{kn} = \frac{1}{h} \sum_r \left[\frac{(\mathbf{e}_\alpha M_{kr})(\mathbf{e}_\beta M_{rn})}{\nu_{rk} - \nu} + \frac{[\alpha \to \beta]}{\nu_{rn} + \nu} \right]. \tag{V, 2-11}$$

It is evident that there are many states R of the radiation field which can cause the molecule to jump from state k to n, and the product $|\, \alpha^{(2)}_{nR} \,|^2$ has to be summed over all the states R. Furthermore, the incident radiation need not be of a single frequency ν_α, and this may also be the case for the emitted photon of frequency ν_β. The number of oscillators of a particular polarization direction j' and propagation vector $\mathbf{k}'$ with frequencies between ν_β and $\nu_\beta + \Delta\nu'$ is given by:

$$\mathrm{d}N^{j'}_\beta = \frac{\nu^2_\beta}{c^3} V \, \mathrm{d}\nu' \, \mathrm{d}\omega' \tag{V, 2-12}$$

where $\mathrm{d}\omega'$ designates a solid angle. The energy density of the radiation field is:

$$\varrho_{j'}(\nu', \omega')\, \mathrm{d}\nu' \, \mathrm{d}\omega' = \sum_\beta^{\Delta\nu' \Delta\omega'} N_\beta h\nu_\beta / V . \tag{V, 2-13}$$

The following expression is now obtained:

$$\sum_R |a^{(2)}_{nR}|^2 = \frac{64\pi^2}{Vh} N_\alpha \cdot \nu_\alpha \sin^2 \Gamma_\alpha \sin^2 \Gamma_\beta \sum_j \iint \left(\varrho_{j'}(\nu', \omega') + \frac{h\nu'^3}{c^3} \right) \times$$

$$\times |M^{\alpha,k'j'}_{kn}|^2 \frac{\sin^2 \pi (\nu_{nk} + \nu_\alpha - \nu')\, t}{(\nu_{nk} + \nu_\alpha - \nu')^2} \mathrm{d}\nu' \, \mathrm{d}\omega' . \tag{V, 2-14}$$

The frequency ν_β in Equation (V, 2-11) is replaced by the variable frequency ν'. The function $[\sin^2 \pi(\nu_{nk}+\nu_\alpha-\nu')t]/(\nu_{nk}+\nu_\alpha-\nu')^2$ has a sharp maximum if $\nu'=\nu_{nk}+\nu_\alpha$ if large time intervals t are taken. It thus appears that only those frequencies which are close to ν' contribute to the integral in Equation (V, 2-14), and assuming that all other terms are constant, the integral:

$$\int_{-\infty}^{+\infty} \frac{\sin^2 \pi(\nu_{nk}+\nu_\alpha-\nu')\,t}{(\nu_{nk}+\nu_\alpha-\nu')^2}\,\mathrm{d}\nu' = \pi^2 t\,. \tag{V, 2-15}$$

Also, $\sin^2 \Gamma_\beta = \frac{1}{2}$ because the phases δ of all vibrators β are independent from each other. Equation (V, 2-14) can thus also be written as:

$$\sum_R |a_{nR}^{(2)}|^2 = \frac{32\pi^4 t}{Vh} N_\alpha \nu_\alpha \sin^2 \Gamma_\alpha \sum_{j'} \left[\varrho_{j'}(\nu'\omega') + \frac{h\nu'^3}{c^3}\right] |M_{kn}^{\alpha,\,k'j'}|\,\mathrm{d}\omega'. \tag{V, 2-16}$$

The energy density of the radiation field due to the incident light waves is given by:

$$\varrho_j(\nu,\omega)\,\mathrm{d}\nu\,\mathrm{d}\omega = \sum_\alpha^{\Delta\nu,\Delta\omega} N_\alpha h\nu_\alpha / V\,. \tag{V, 2-17}$$

If W_{nk} is the probability per unit time for the transition $k \to n$ then, by substituting Equation (V, 2-17) in Equation (V, 2-16):

$$\begin{aligned} W_{nk}(\nu,\nu') = \frac{16\pi^4}{h^2} \sum_{jj'} \iiint \varrho_j(\nu,\omega)\,\times \\ \times \left[\varrho_{j'}(\nu',\omega') + \frac{h\nu'^3}{c^3}\right] |M_{nk}^{\mathbf{k}j,\mathbf{k}'j'}|^2\,\mathrm{d}\nu\,\mathrm{d}\omega\,\mathrm{d}\omega' \end{aligned} \tag{V, 2-18}$$

where j is the direction of polarization of the incident light wave and $\mathbf{k}$ is the wave vector. A photon of energy $h\nu$ is thus destructed in favour of a photon of energy $h\nu'$ which is created, where $\nu'=\nu+\nu_{nk}$. An interesting point is the appearance of a radiation field of density $\varrho(\nu'=\nu+\nu_{nk})$. It may be recalled here that the rate of induced emission of a normal absorption-emission process is proportional to the density of the radiation for field with a frequency equal to that of the actual transition Equation (I, 3-3). Also, the ratio of the rate of spontaneous and induced emission is given by $(8\pi h\nu/c^3)$ $[1/\varrho(\nu)]$. This follows from Equation (I, 3-1), (I, 3-3) and (I, 3-7). Apparently, induced and spontaneous processes are also present in the creation of photons of energy $h(\nu+\nu_{kn})$ because an identical ratio occurs. However, it is well to realize that for the Raman process there does not exist an excited state from which emission occurs. The Equation (V, 2-18) yields only the results that photons are destructed and created, and consequently, it is not appropriate to discuss the increase of rate of stimulated over spontaneous emission in terms of population inversion of an excited state as is the case in the normal emission process Equation (I, 3-11). The information which is obtained from stimulated Raman scattering is limited if the results are applied to structural molecular spectroscopy. Not all Raman lines which comprise the spectrum of a molecule can be made to become 'stimulated'. In general, only the strongest lines of the spectrum are good candidates. The intensity of such a line may become so strong that the Raman shift assumes the role of exciting line and so on. This has been observed for benzene $\Delta\nu_1 = 992$ cm^{-1} and shifts of $\Delta\nu = 3\times 992$ cm^{-1} have been observed. In other cases, apparent anti-Stokes stimulated Raman lines were observed. But the explanation of the observation is as follows. Two incident photons are destructed, and the created photons are of frequency $\nu_0 - \nu_{kn}$ and $\nu_0 + \nu_{kn}$ so that conservation of energy has been achieved because $2h\nu_0 = h\nu_0 - h\nu_{kn} + h\nu_0 + h\nu_{kn}$. The distribution of the stimulated radiation in space is quite different from that of the normal Raman scattering. This particular property of the stimulated Raman effect can be traced back to the condition that – apart from energy – conservation of wave vectors should also be maintained. The result is that, under certain conditions, the emitted radiation occurs in rings around the forward scattered laser beam.

Stimulated Raman scattering is mostly achieved if giant pulse lasers are used. The pulse time of such a laser is in the order of 10^{-7}–10^{-8} s, and at the end of this pulse, the first vibrational excited state is populated to a high degree. For H_2, for instance, the first vibrational level at 4156 cm^{-1} can be populated. By making an observation of the decrease of the Raman intensity as a function of time, the relaxation time of this state could be determined.

The observation of the stimulated Raman effect in 1962 by Woodbury and Ng has aroused the interest of many physicists, and the theory of this effect has been well worked out after the discovery of this type of scattering in the early sixties. In the ensuing excitement, it occurs to this author that the work of Placzek has been overlooked to a certain extent.

V-3. Induced Absorption: The Inverse Raman Effect

The polarization properties of spontaneous and stimulated Raman scattering are closely related to the matrix elements $M_{nk}^{\mathbf{k}j;\,\mathbf{k}'j'}$ of Equation (V, 3-18). Thus:

$$|M_{kn}^{\mathbf{k}j,\mathbf{k}'j'}|^2 = \frac{1}{h^2}\int \sum_{jj}\sum_{r}\left[\frac{(\mathbf{e}_{kj}\mathbf{M}_{kr})\,(\mathbf{e}_{\mathbf{k}'j'}\mathbf{M}_{rn})}{\nu_{rk}-\nu} + \frac{(\mathbf{e}_{\mathbf{k}'j'}\mathbf{M}_{kr})\,(\mathbf{e}_{kj}\mathbf{M}_{rn})}{\nu_{rn}+\nu}\right] \tag{V, 3-1}$$

where $\mathbf{e}_{kj}$ is a unit vector, j is the direction of polarization and k is the direction of propagation ($\mathbf{e}_{kj}$ is perpendicular to k). $\mathbf{M}_{kr}$ and $\mathbf{M}_{rn}$ have the usual meaning. This expression can be simplified if the incident radiation is a plane monochromatic wave, thus:

$$\mathbf{E} = \mathbf{A}^* e^{2\pi i\nu t} + \mathbf{A}e^{-2\pi i\nu t} \tag{V, 3-2a}$$

and

$$\iint \bar{\varrho}_j(\nu, \omega)\,\mathrm{d}\nu\,\mathrm{d}\omega = \frac{\overline{\mathbf{E}^2}}{4\pi} = \frac{|\mathbf{A}|^2}{2\pi} \tag{V, 3-2b}$$

because $\varrho(\nu, \omega) \equiv 0$ for monochromatic radiation. If Equation (V, 3-2b) is substituted into Equation (V, 3-1), then the result is:

$$(M_{kn})^{\mathbf{k}j;0} = \frac{|\mathbf{A}|^2}{4\pi} \int \frac{1}{h^2} \sum_{j'} \left| \sum_r \left[\frac{(\mathbf{e}_0 \mathbf{M}_{kr})(\mathbf{e}_{\mathbf{k}'j'} \mathbf{M}_{rn})}{\nu_{rk} - \nu} + \frac{(\mathbf{e}_{\mathbf{k}'j'} \mathbf{M}_{kr})(\mathbf{e}_0 \mathbf{M}_{rn})}{\nu_{rn} + \nu} \right] \right|^2 \tag{V, 3-3}$$

where $\mathbf{e}_0$ is the unit vector parallel to the direction of polarization of the incident wave. In Equation (V, 3-3), products of the type $\int \sum_j |\, \mathbf{e}_{\mathbf{k}'j'} \mathbf{P} \,|^2 \, d\omega'$ occur. Radiation is fixed by two states of polarization ($j' = 1, 2$) so that:

$$\int \sum_{j'} |\mathbf{e}_{\mathbf{k}'j'} \mathbf{P}|^2 \, d\omega' = \int |\mathbf{e}_{\mathbf{k}'1} \mathbf{P}|^2 + |\mathbf{e}_{\mathbf{k}'2} \mathbf{P}|^2 \, d\omega' . \tag{V, 3-4}$$

If the directions of polarization are so chosen that $\mathbf{e}_{\mathbf{k}1}$ is in the plane formed by the direction of propagation $\mathbf{k}^1$ and $\mathbf{P}$, then $\mathbf{e}_{\mathbf{k}2}$ is perpendicular to this plane; then (V, 3-4) reduces to

$$\int \sum_{j'} |\mathbf{e}_{k'j'} \mathbf{P}|^2 \, d\omega' = \int |\mathbf{e}_{\mathbf{k}1} \mathbf{P}|^2 \, d\omega'$$
$$= \frac{8\pi}{3} |\mathbf{P}|^2 .$$

Thus:

$$|\mathbf{M}^0_{kn}|^2 = \frac{8}{3} |\mathbf{A}^2| \left| \sum_r \left[\frac{(\mathbf{e}_0 \mathbf{M}_{rk})(\mathbf{M}_{rn})}{\nu_{rk} - \nu} + \frac{(\mathbf{M}_{rk})(\mathbf{e}_0 \mathbf{M}_{rn})}{\nu_{rn} + \nu} \right] \right|^2 . \tag{V, 3-5}$$

The transition probability $n \leftarrow k$ per unit time can now be written as:

$$W_{kn}(\nu, \nu + \nu_{nk}) = \frac{8\pi^4 (\nu + \nu_{nk})^3}{3hc^3} |\mathbf{M}^0_{kn}|^2 \tag{V, 3-6}$$

and the scattered intensity of frequency $\nu + \nu_{nk}$ is:

$$I_{nk} = h(\nu + \nu_{nk}) \, W_{nk} . \tag{V, 3-7}$$

Equation (V, 3-5) represents the intensity emitted by the classical dipole moment $\mathbf{M} = \mathbf{C}_{nk} e^{-2\pi i(\nu + \nu_{nk})t} + \mathbf{C}^*_{nk} e^{2\pi i(\nu + \nu_{nk})t}$ where

$$\mathbf{C}_{nk} = \frac{1}{h} \sum_r \left[\frac{(\mathbf{A}\mathbf{M}_{kr})(\mathbf{M}_{rn})}{\nu_{rk} - \nu} + \frac{\mathbf{M}_{kr}(\mathbf{A}\mathbf{M}_{rn})}{\nu_{rn} + \nu} \right] . \tag{V, 3-8}$$

The results of the correspondence and of the radiation field theories are thus identical.

If the densities $\varrho_{j'}(\nu', \omega')$ and $\varrho_j(\nu, \omega)$ are different from zero but independent of $j_1^1\omega^1$ and j, ω respectively, then the incident and emitted radiation is spatially isotropic and not quite monochromatic.

Now

$$\varrho(\nu) = \sum_j \int \varrho_j(\nu, \omega)\, d\omega = 8\pi\varrho_j(\nu, \omega)$$

and

$$\varrho(\nu') = 8\pi\varrho_{j'}(\nu', \omega'). \qquad \text{(V, 3-9)}$$

The frequency distribution of $\varrho(\nu)$ is so that $\varrho(\nu) \neq 0$ in the range of $\Delta\nu$, but $\varrho(\nu)=0$ outside this small region. Thus, $\int \varrho(\nu)d\nu = \bar{\varrho}(\nu)\Delta\nu$ and, after integration over $d\omega$, $d\omega^1$, the expression for W_{nk} becomes:

$$W_{nk}(\nu, \nu') = \bar{\varrho}(\nu_0)\left[\frac{8\pi\nu'^3}{c^3} + \varrho(\nu')\right]|\mathbf{M}_{nk}|^2\, \Delta\nu \qquad \text{(V, 3-10)}$$

where

$$|\mathbf{M}_{nk}|^2 = \sum_{\varrho}\sum_{\varrho'} \frac{16\pi^4}{3h^2}\left|\sum_r\left[\frac{(\mathbf{M}_\varrho)_{kr}(\mathbf{M}_{\varrho'})_{rn}}{\nu_{rk}-\nu} + \frac{(M_{\varrho'})_{kr}(M_\varrho)_{rn}}{\nu_{rn}+\nu}\right]\right|. \qquad \text{(V, 3-10a)}$$

For non-spatially isotropic emitted and incident radiation, the expression for W_{nk} is:

$$W_{nk} = \int |\mathbf{M}_{nk}|^2\, \varrho(\nu)\left[\varrho(\nu' = \nu + \nu_{kn}) + \frac{8\pi^3 h(\nu + \nu_{kn})^3}{c^3}\right] \qquad \text{(V, 3-11)}$$

where $\mathbf{M}_{nk} = \alpha_{\varrho\sigma}\mathbf{E}_\sigma$ and $\alpha_{\varrho\sigma}$ has its usual meaning. The product $\varrho(\nu)\,\varrho(\nu+\nu_{kn})$ is related to an induced absorption process of a light quantum of energy $h\nu$ followed by the induced emission of a photon of energy $h(\nu+\nu_{kn})$. In discussing Raman scattering, it is not quite correct to speak of absorption and emission if the absorption is between the ground state and an excited state, and the emission is from this excited state to another lower-lying state. Rather, the scattering process can be described as the interaction of a photon and the molecule, which interaction results in the emission of a photon of different energy while the molecule undergoes a change in energy. Because of the close correspondence of the classical picture of absorption and emission and the actual process of light scattering, the words 'absorption' and 'emission' are sometimes used. In the normal spontaneous Raman effect, the term $\varrho(\nu)\, 8\pi^3 h(\nu + \nu_{kn})^3/c^3$ dominates over the stimulated effect $\varrho(\nu)\,\varrho(\nu+\nu_{kn})$.

However, if the molecule is simultaneous irradiated with two intense light beams of frequency ν_0 and $\nu_0+\nu_{kn}$, then an inverse scattering process can take place. Under the influence of the radiation field with the densities $\varrho(\nu_0+\nu_{kn})$ and $\varrho(\nu_0)$, molecules in the quantum state k are induced to simultaneously absorb radiation at $\nu_0+\nu_{kn}$ and emit radiation at ν_0. Thus, if the spectral region at $\nu=\nu_0+\nu_{kn}$ is studied, then during the scattering process an absorption at that frequency should be observed. This inverse Raman effect which was first observed by Stoicheff and Jones in 1964 is also depicted in Figure V-2.

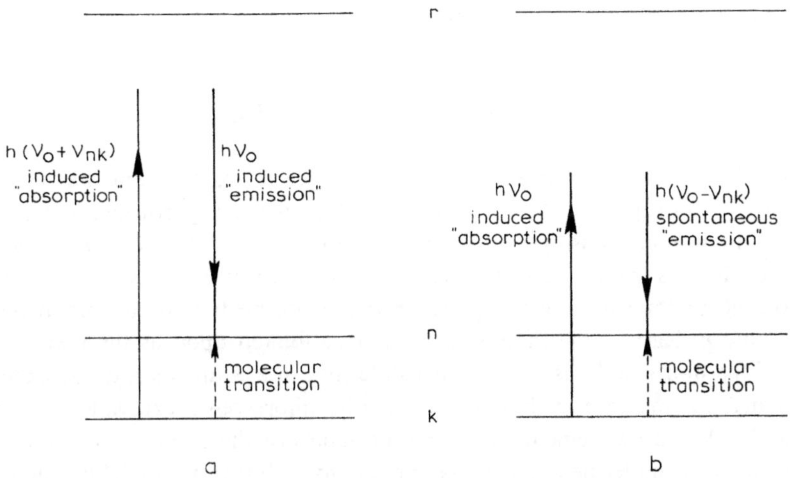

Fig. V-2. A comparison of the inverse (a) and spontaneous (b) Stokes Raman effects.

V-4. The Resonance Raman Effect

Probably the least understood scattering process is the case where the frequency of the incident radiation is close to or coincident with the frequency of an absorption band of the molecule, ion or atom. The quantum mechanical perturbation calculation for wave functions (and thus also for an element of the scattering) is not valid any more because the perturbation is now large. The value for an element of the scattering tensor for the non-resonance case is:

$$(\alpha_{\varrho\sigma})_{kn} = \frac{1}{h}\sum_r \left[\frac{(M_\sigma)_{kr}(M_\varrho)_{rn}}{\nu_{rk} - \nu_0} + \frac{[\sigma \leftrightarrow \varrho]}{\nu_{rn} + \nu_0}\right] \qquad \text{(V, 4-1)}$$

and it is seen that $(\alpha_{\varrho\sigma})_{kn}$ becomes large if $\nu_{nk} - \nu_0 \approx 0$. Thus Equation (V, 4-1) should, in fact, not be used because perturbation calculation of any property is not permitted to yield large values of such properties. Nevertheless, this expression is used to explain certain features of near-resonance and resonance Raman spectra.

It is clear that the absorption band of the molecule plays an important role in the intensity of Raman bands in the resonance-Raman effect. Weiskopf introduced the width of the absorption band in the expression for α in the following way. If τ_r is the lifetime of the excited state r which is in resonance with ν_0, then a damping factor γ_r can be defined where:

$$\gamma_r = \frac{1}{4\pi\tau_r}. \qquad \text{(V, 4-2)}$$

In the final analysis, this damping factor γ_r appears as an additional imaginary quantity

$i\gamma_r$ in the denominators of the expression (V, 4-1) in the non-resonance case. Thus:

$$(\alpha_{\varrho\sigma})_{kn} = \frac{1}{h}\sum_r \left[\frac{(M_\sigma)_{kr}(M_\varrho)_{rn}}{\nu_{rk} - \nu_0 + i\gamma_r} + \frac{[\varrho \leftrightarrow \sigma]}{\nu_{rn} + \nu_0 + i\gamma_r}\right]. \qquad \text{(V, 4-3)}$$

It should be pointed out here that the appearance of excited state wave functions – via $(M_\varrho)_{kr}$ – in Equation (V, 4-3) is a result of a particular perturbation calculation, and that the Raman transition $n \leftarrow k$ is *not* represented by a series of transitions where the molecule goes from $r \leftarrow k$ and $n \leftarrow r$. Consequently, it is *not* possible to believe that the molecule remains for a certain period τ_r in the intermediate state r and then returns to the state n. Rather, the lifetime τ_r should be looked upon as the lifetime of the state r if the molecule finds itself in this state (not in a Raman scattering experiment but during, for instance, a fluorescence or phosphorescence experiment). Also, the electric dipole matrix elements **M** represent values of the strength of electric dipole transitions if the molecule could actually undergo such transitions. Thus, the strength of absorption and fluorescence (phosphorescence) bands may be used to investigate which of the matrix elements $(M)_{rk}$ contributes to the expression of α, but during the scattering process, the molecule goes from $k \leftrightarrow n$.

It is always possible to find a spectral region of the electromagnetic spectrum of the molecule which is sufficiently far away from resonance so that the damping term $i\gamma_r$ can be neglected, but sufficiently close so that the summation over the excited states r can be restricted to the (vibronic) levels of one electronic (the resonating) excited state r. Because of the emphasis placed on this state r, which is in near resonance, the second member of (V, 4-3) can be neglected and:

$$(\alpha_{\varrho\sigma})_{kn} \approx \frac{1}{h}\sum_r \frac{(M_\sigma)_{kr}(M_\varrho)_{rn}}{\nu_{rk} - \nu_0}. \qquad \text{(V, 4-4)}$$

Placzek's polarizability theory for the normal vibrational and rotational Raman effect employs the polarizability $(\alpha)_{kk}$ of the ground state, thus:

$$(\alpha_{\varrho\sigma})_{kk} \approx \frac{1}{h}\sum_r \frac{(M_\sigma)_{kr}(M_\varrho)_{rk}}{\nu_{rk} - \nu_0}$$

and

$$(\alpha_{\varrho\varrho})_{kk} \approx \frac{1}{h}\sum_r \frac{|(M_\varrho)_{kr}|^2}{\nu_{rk} - \nu_0}. \qquad \text{(V, 4-5)}$$

Differentiation to the normal coordinate Q_m yields the result:

$$\frac{\partial(\alpha_{\varrho\varrho})_{kk}}{\partial Q_m} = \frac{1}{h}\sum_r \left\{\frac{2\nu_{rk}}{\nu_{rk}^2 - \nu_0^2}\frac{\partial|(M_\varrho)_{kr}|^2}{\partial Q_m} - \frac{2(\nu_{rk}^2 + \nu_0^2)}{(\nu_{rk}^2 - \nu_0^2)^2}|(M_\varrho)_{kr}|^2\frac{\partial \nu_{rk}}{\partial Q_m}\right\}. \qquad \text{(V, 4-6)}$$

It has been demonstrated by Behringer and Brandmüller that in the near-resonance – but also at resonance – the second term dominates over the first, and thus the intensity I_{kn} is given by:

$$I_{kn} \sim (\nu_0 - \nu_{kn})^4 \frac{\nu_{rk}^2 + \nu_0^2}{\nu_{rk}^2 - \nu_0^2}. \tag{V, 4-7}$$

This is the expression derived by Shorygin, and it may be noted that the composition of the resonating state r is restricted in that only the pure electronic transitions $r \leftarrow k$ are considered. It is well known that vibrational transitions appear with the pure electronic transitions and, using the notation of chapter IV, such transitions may be designated by $k, v_m \rightarrow r, v''_m$, where v_m and v''_m are vibrational quantum numbers. The following expression can be derived for the resonance Raman effect if the vibronic coupling model introduced by Albrecht is used:

$$\begin{aligned}(\alpha_{\varrho\sigma})_{k,\, v_m;\, kv'_m} = {} & \frac{1}{h} \sum_{r,\, v''_m} \left[\frac{(M_\sigma^0)_{kr} (M_\varrho^0)_{rk}}{\nu_{r,\, v''_m;\, k,\, v_m} - \nu_0 + i\gamma_{r,\, v''_m}} \times \right. \\ & \left. \times \langle \varphi_{v_m}(\mathbf{Q}_m) \mid \varphi_{v''_m}(\mathbf{Q}_m)\rangle \langle \varphi_{v''_m}(\mathbf{Q}_m) \mid \varphi_{v'_m}(\mathbf{Q})\rangle \right] + \\ & \sum_{r,\, t,\, v''_m} \frac{\begin{aligned} h_{tr}^{Q_m} [(M_\sigma^0)_{kt} (M_\varrho^0)_{rk} \langle \varphi_{v''_m}(\mathbf{Q}_m) \mid \varphi_{v'_m}(\mathbf{Q}_m)\rangle \langle \varphi_{v_m}(\mathbf{Q}_m) | \, \mathbf{Q}_m \, | \varphi_{v''_m}(\mathbf{Q}_m)\rangle + \\ + (M_\varrho^0)_{tk} (M_\sigma^0)_{kr} \langle \varphi_{v_m}(\mathbf{Q}_m) \mid \varphi_{v''_m}(\mathbf{Q}_m)\rangle \langle \varphi_{v''_m}(\mathbf{Q}_m) | \, \mathbf{Q}_m \, | \varphi_{v'_m}(Q_m)\rangle \end{aligned}}{\nu_{r,\, v''_m;\, k,\, v_m} - \nu_0 + i\gamma_{r,\, v''_m}} + \\ & + \sum_{\substack{r,\, t,\, v''_m \\ (t \neq k)}} \frac{\begin{aligned} h_k^{Q_m} [(M_\varrho^0)_{rk} (M_\sigma^0)_r \langle \varphi_{v'_m}(\mathbf{Q}_m) \mid \varphi_{v''_m}(\mathbf{Q}_m\rangle \langle \varphi_{v''_m}(\mathbf{Q}_m) | \, Q \, | \varphi_{v_m}(Q_m)\rangle + \\ + (M_\sigma^0)_{kr} (M_\varrho^0)_{tr} \langle \varphi_{v_m}(\mathbf{Q}_m) \mid \varphi_{v''_m}(\mathbf{Q}_m)\rangle \langle \varphi_{v''_m}(Q_m) | \, Q_m \, | \varphi_{v'_m}(\mathbf{Q}_m)\rangle \end{aligned}}{v_{r,\, v''_m;\, k,\, v_m} - \nu_0 + i\, \gamma_{r,\, v''_m}}\end{aligned} \tag{V, 4-8}$$

This expression is derived by substitution of Equations (IV, 1-10a) and (IV, 1-11a) into Equation (IV, 1-12) and by letting the electronic quantum number $n = k$. It is obvious that as a result of resonance of ν_0 and $\nu_{r,\, v''_m;\, k,\, v_m}$ the summation $\sum_{v''_m}$ cannot be carried out. Equation (V, 4-8) is valid in the case where the wave functions of the ground state k and the resonating state r are admixed with the wave functions t. At the heart of this admixture lies the effect of vibronic coupling of one particular normal coordinate $\mathbf{Q}_m$ of the molecule with all electronic states.

As a consequence of the fact that the closure rule cannot be applied, the integrals $\langle \varphi_{v_m}(\mathbf{Q}_m) \mid \varphi_{v''_m}(\mathbf{Q}_m)\rangle \langle \varphi_{v''_m}(\mathbf{Q}_m \mid \varphi_{v'_m}(\mathbf{Q}_m)\rangle$; $\langle \varphi_{v''_m}(\mathbf{Q}_m) \mid \varphi_{v'_m}(\mathbf{Q}_m)\rangle \langle \varphi_{v''_m}(\mathbf{Q}_m) \times$ $\times |\mathbf{Q}_m \mid \varphi_{v_m}(\mathbf{Q}_m)\rangle$; $\langle \varphi_{v_m}(\mathbf{Q}_m) \mid \varphi_{v''_m}(\mathbf{Q}_m)\rangle \langle \varphi_{v''_m}(\mathbf{Q}_m) \mid \mathbf{Q}_m \mid \varphi_{v'_m}(\mathbf{Q}_m)\rangle$ and the corresponding denominators $\nu_{r,\, v''_m;\, k v,_m} - \nu_0 + i\gamma_r$ should be computed for individual values of v''_m, v'_m and v_m. In Table V-1, selection rules are given as derived from the three integrals, assuming that of all excited states of the molecule, only the contribution of the resonating vibronic state r, v''_m to $(\alpha_{\varrho\sigma})_{k,\, v_m;\, k,\, v'_m}$ is important. The first term of Equation (V, 4-8) represents Rayleigh radiation in the non-resonance case, but can

TABLE V-1

Selection rules for the transition $v_m \overset{\Delta v_m}{\leftrightarrow} v'_m$ for resonance Raman scattering

Condition of resonance	Condition of non-vanishing of the integrals			
$\nu_{r,v''_m;k,v_m} - \nu_0 \approx 0$		$\langle \varphi_{v_m}(\mathbf{Q}_m) \vert \varphi_{v''_m}(\mathbf{Q}_m) \rangle \times \langle \varphi_{v''_m}(\mathbf{Q}_m) \vert \varphi_{v'_m}(\mathbf{Q}_m) \rangle$	$\langle \varphi_{v_m}(\mathbf{Q}_m) \vert \varphi_{v''_m}(\mathbf{Q}_m) \rangle \times \langle \varphi_{v''_m}(\mathbf{Q}_m) \vert \mathbf{Q}_m \vert \varphi_{v'_m}(\mathbf{Q}_m) \rangle$	$\langle \varphi_{v'_m}(\mathbf{Q}_m) \vert \varphi_{v''_m}(\mathbf{Q}_m) \rangle \times \langle \varphi_{v''_m}(\mathbf{Q}_m) \vert \mathbf{Q}_m \vert \varphi_{v_m}(\mathbf{Q}_m) \rangle$
$v''_m = v_m$	(a)	$\Delta v_m = 0$	$\Delta v_m = \pm 1$	integral continuously zero
	(b)	$\Delta v_m = 0, 1, 2, 3 \ldots$	$\Delta v_m = 0, \pm 1, \pm 2 \ldots$	$\Delta v_m = 0, \pm 1, \pm 2 \ldots$
$v''_m = v_m + 1$	(a)	integral continuously zero	integral continuously zero	integral continuously zero
	(b)	$\Delta v_m = 0, \pm 1, \pm 2, \pm 3 \ldots$	$\Delta v_m = 0, \pm 1, \pm 2$	$\Delta v_m = 0, \pm 1, \pm 2 \ldots$
$v''_m = v_m + n$	(a)	integral continuously zero	integral continuously zero	integral continuously zero
	(b)	$\Delta v_m = 0, \pm 1, \pm 2, \pm 3, \ldots n, \ldots$	$\Delta v_m = 0, \pm 1, \pm 2, \ldots n, \ldots$	$\Delta v = 0, \pm 1, \pm 2 \ldots n, \ldots$

(a) The potential surfaces of the electronic states k and r are equal.
(b) Not equal.

If the potential surfaces are approximately equal then the intensity of the transitions for which $\Delta v_m = n$ is the strongest of the series if $\nu_0 \approx \nu_{r,v''_m;k,v'_m}$ where $\nu_{''m} = \nu_m + n$.

contribute to Raman scattering in the resonance case if the potential surfaces of the states k and r differ from each other. The appearance of overtones in the Raman spectrum can also be related to the difference of these surfaces. If resonance occurs with the nth vibronic state or r, then the nth overtone of the ground state should show a maximum intensity if compared to the intensity of the other overtones if the potential surfaces of ground and excited states are equal. The polarization properties of the Raman lines can also be different because of the occurrence of trace scattering and asymmetric tensor; but antisymmetric tensors cannot be present.

Behringer's approach to resonance Raman scattering originates from expression (IV, 4-3a). If the damping factor γ_r is introduced and resonance exists, then:

$$(\alpha_{\varrho\sigma})_{kv_m;\,kv'_m} \approx \frac{1}{h} \sum_{v''_m} \frac{(M^0_\sigma)_{kr}\,(M^0_\varrho)_{rk}}{\nu_{r,\,v''_m;\,k,\,v_m} - \nu_0 + i\gamma_r} \times \\ \times \langle \varphi_{v_m}(\mathbf{Q}_m) | \, \varphi_{v''_m}(\mathbf{Q}_m) \, \langle \varphi_{v''_m}(\mathbf{Q}_m) \, | \, \varphi_{v'_m}(\mathbf{Q}_m) \rangle . \qquad \text{(V, 4-9)}$$

The summation over all states rv''_m is reduced to a single summation over all vibrational levels v''_m of the resonating state r, and the frequency $v_{r''m;\,kv_m}$ is again in the vicinity of v_0. The condition of non-vanishing of the integral $\langle \varphi_{v_m}(\mathbf{Q}_m) \mid \varphi_{v''_m}(\mathbf{Q}_m)\rangle$ $\langle \varphi_{v''_m}(\mathbf{Q}_m \mid \varphi_{v'_m}(\mathbf{Q}_m)\rangle$ has been discussed in the above, and the results are given in Table V-1. From a consideration of the product matrix elements involving the vibrational wave functions $\varphi_{v_m}(\mathbf{Q}_m)$ etc., it may be shown that for a specific difference of the potential functions, the selection rule $\Delta v_m = \pm 1$ dominates. Behringer introduced a rectangular shape for the absorption band and showed that the scattering tensor can be decomposed in a real α_r and imaginary α_i part as follows:

$$\alpha = \alpha_r + \alpha_i \qquad \text{(V, 4-10)}$$

and

$$I \approx (\alpha_r)^2 + (\alpha_i)^2 . \qquad \text{(V, 4-10a)}$$

The behaviour of α and I is also shown in Figure V-3 where ν_0 changes smoothly through the region of absorption.

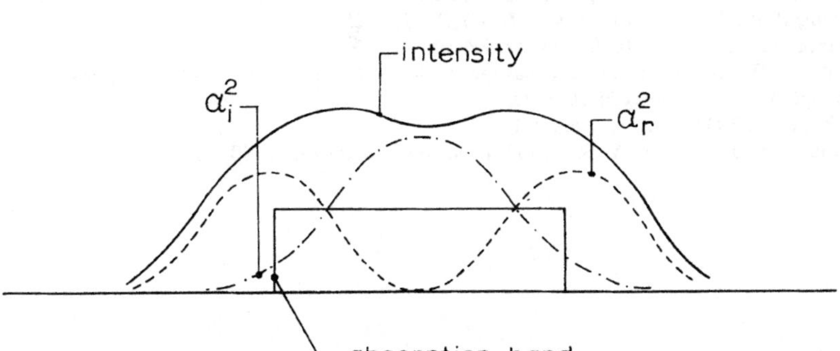

Fig. V-3. The effect of damping on the intensity of a band (resonance Raman).

Not only the position but also the symmetry of the excited resonating state r should be considered. Assuming that the molecule belongs to the point group D_2, all states should belong to one of the representations A_1, B_1, B_2 or B_3 of this point group. If the ground state transforms according to the totally symmetric species, and the resonating state r is of symmetry B_1, then the element α_{xx} of the scattering tensor is increased in value, and only the totally symmetric vibrations of this molecule show an enhanced intensity. This follows from the first member of Equations (V, 4-8) which in fact corresponds to Behringer's expression. The second and third member of Equation (V, 4-7) yields the result that the enhanced elements of the scattering tensor are α_{xx}, α_{yx} and α_{xz}. From the conditions of non-vanishing of the constants $h_{tk}^{Q_m}$ and $h_{tr}^{Q_m}$, it follows that for a A_1 mode, α_{xx} is the only element which is enhanced (but α_{yy} and α_{zz} are not equal to zero because of the small contribution to α of all other non-resonating states). For a normal mode of symmetry B_1, the tensor element α_{xy} is enhanced, and not α_{yx}. Thus, the tensor becomes asymmetric and for a system of random oriented particles $\frac{6}{7} < \varrho_n < 2$. For a mode of symmetry B_3, it is again found that $\frac{6}{7} < \varrho_n < 2$, but a normal mode of symmetry B_2 is not enhanced, and $\varrho_n = \frac{6}{7}$ and $\varrho_l = \frac{3}{4}$. From a theoretical point of view, it appears that Albrecht's model is more flexible than the quantum mechanical model of Behringer. Also, in the final analysis, it is true that Albrecht and Shorygin predict the same frequency dependency for the intensity of Raman bands if resonance conditions exist.

Suggestions for Further Reading

Albrecht, A. C.: 1961, *J. Chem. Phys.* **34**, 1476.
Behringer, J. and Brandmuller, J.: 1956, *Z. Elektrochem.* **60**, 643.
Bloembergen, N.: 1967, *Am. J. Phys.* **35**, 989.
Christie, J. H. and Lockwood, D. J.: 1971, *J. Chem. Phys.* **54**, 1141.
Cyvin, S. J., Rauch, J. E., and Decius, J. C.: 1965, *J. Chem. Phys.* **43**, 4083.
Jones, W. J. and Stoicheff, B. P.: 1964, *Phys. Rev. Letters* **13**, 657.
Long, D. A. and Stanton, L.: 1970, *Proc. Roy. Soc. (London)* **A 318**, 441.
Placzek, G.: 1934, *Marx Handbuch der Radiologie* (2d ed.), Vol. VI, Academische Verlagsgesellschaft, Leipzig, p. 209.
Rivoire, G. and Dupeyrat, R.: 1963, *Compt. Rend. Acad. Sci.* **256**, 1947.
Rokni, M. and Yatsiv, S.: 1967, *Phys. Letters* **24A**, 277.
Shorygin, P. P.: 1947, *J. Phys. Chem. U.S.S.R.* **21**, 1125.
Sorokin, P. P. *et al.*: 1967, *Appl. Phys. Letters* **10**, 44.
Terhune, R. W., Maker, P. D., and Savage, C. M.: 1965, *Phys. Rev. Letters* **14**, 681.
Weisskopf, V.: 1931, *Ann. Physik* **9**, 23.
Weisskopf, V.: 1933, *Z. Physik* **85**, 451.
Woodbury, W. J. and Ng, W. K.: 1967, *Proc. Inst. Radio Engrs.* **50**, 2367.

APPENDIX I

PROPERTIES OF REPRESENTATIONS OF SOME IMPORTANT POINT GROUPS

Throughout the last three chapters of this book, use has been made of the character tables of some point groups, and in particular the transformation properties of the scattering tensor were discussed. These properties together with the characters and the species of some of the more important point groups are given in Tables A-I–A-X.

In the first column of the table is given the symbols of the irreducible representations as proposed by Mulliken. All one-dimensional representations are designated by the letters A and B, the two-dimensional representations by E and the three-dimensional species by F (or T).

The one-dimensional species which are symmetric with respect to rotation by $2\pi/n$ about the principal C_n axis are represented by the letter A, and those who are antisymmetric with respect to the rotation are designated B. If such a principal axis is perpendicular to a plane, or if a C_2 axis exists which is perpendicular to the principal axis, then the subscripts 1 and 2 are used to distinguish the species. The subscript 1 is used to indicate that the species is symmetric with respect to a reflection through the plane or a rotation of π degrees, and the subscript 2 is attached to the species to indicate that this species is antisymmetric to these operations.

Also, primes and double primes are attached to the species of some groups to indicate which one is symmetric or antisymmetric with respect to a reflection by a horizontal mirror plan. If one of the symmetry elements is the center of inversion, then the subscript g (gerade) is attached to all representations which are symmetric, and the subscript u (ungerade) to those representations which are antisymmetric to the inversion center.

For point groups with an inversion center, the character tables are not given explicitly, but can be derived from the corresponding groups without a center of inversion by replacing each species by two: one symmetric (g) and one antisymmetric (u) with respect to the center of inversion. The group has twice the number of species and symmetry elements if compared to the corresponding group without i. The above is also given by relations of the type $O_h = O \times C_i$.

Symmetry elements are given directly above the characters of the groups. The symmetry element R is an artificial one which occurs only in extended or double groups. For these groups the order of classes of two-fold axes and of planes is double the number shown in the tables. The species of the double groups are given below the broken line and the characters are on the right of the vertical broken line.

In the first column under the heading 'Raman activity' are the elements of the scattering tensor and in the second column are the elements of the irreducible tensor.

TABLE A-I

Point Groups C_1, C_2, C_3, C_s, C_i

C_1	E	R	Raman activity			Dipole moment
A	1	1	all $\alpha_{\varrho\sigma}$'s	all α_Q^K's	all $\beta_{\varrho\sigma\sigma'}$	x, y, z
$B_{1/2}$	1	-1				

C_2	E	$C_2(z)$	R	Raman activity			Dipole moment
A	1	1	1	$\alpha_{xx}, \alpha_{yy}, \alpha_{zz}, \alpha_{xy} \pm \alpha_{yx}$	$\alpha_0^0, \alpha_0^1, \alpha_2^2, \alpha_0^2, \alpha_{-2}^2$	$\beta_{zzz}, \beta_{yyz}, \beta_{zxx}, \beta_{xyz}$	z
B	1	-1	1	$\alpha_{zx} \pm \alpha_{xz}, \alpha_{zy} \pm \alpha_{yz}$	$\alpha_1^1, \alpha_{-1}^1, \alpha_1^2, \alpha_{-1}^2$	$\beta_{xxx}, \beta_{yyy}, \beta_{xxy}, \beta_{zzx}, \beta_{xyy}, \beta_{yzz}$	x, y
$E_{1/2}$	2	0	-2				

C_3	E	$2C_3$	R	$2C_3^2$	Raman activity			Dipole moment
A	1	1	1	1	$\alpha_{xx} + \alpha_{yy}, \alpha_{zz}, \alpha_{xy} - \alpha_{yx}$	$\alpha_0^0, \alpha_0^1, \alpha_0^2$,	$\beta_{xxx} - 3\beta_{xyy}, \beta_{yyy} - 3\beta_{xxy}, \beta_{yyz}+ +\beta_{zxx}, \beta_{zzz}$	z
E	2	-1	2	-1	$(\alpha_{zx} \pm \alpha_{xz}, \alpha_{yz} \pm \alpha_{zy})$, $(\alpha_{xx} - \alpha_{yy}, \alpha_{xy} + \alpha_{yx})$	$\alpha_1^1, \alpha_{-1}^1, \alpha_2^2, \alpha_1^2, \alpha_{-1}^2, \alpha_{-2}^2$	$(\beta_{xxx} + \beta_{xyy}, \beta_{yyy} + \beta_{xxy}), (\beta_{zzx}, \beta_{yzz})$, $(\beta_{yyz} - \beta_{zxx}, \beta_{xyz})$	(x, y)
$E_{1/2}$	2	1	-2	-1				
$B_{3/2}$	1	-1	-1	1				

C_s	E	$\sigma(xy)$	R	Raman activity			Dipole moment
A'	1	1	1	$\alpha_{xx}, \alpha_{yy}, \alpha_{zz}, \alpha_{xy} \pm \alpha_{yx}$	$\alpha_0^0, \alpha_0^1, \alpha_2^2, \alpha_0^2, \alpha_{-2}^2$	$\beta_{xxx}, \beta_{yyy}, \beta_{xxy}, \beta_{zzx}, \beta_{xyy}, \beta_{yzz}$	x, y
A''	1	-1	1	$\alpha_{zx} + \alpha_{xz}, \alpha_{yz} + \alpha_{zy}$	$\alpha_1^1, \alpha_{-1}^1, \alpha_1^2, \alpha_{-1}^2$	$\beta_{zzz}, \beta_{yyz}, \beta_{zxx}, \beta_{xyz}$	z
$E_{1/2}$	2	0	-2				

C_i	E	i	R	iR	Raman activity		Dipole moment
A_g	1	1	1	1	all $\alpha_{\varrho\sigma}$'s	all α_Q^K's	
A_u	1	-1	1	-1		all $\beta_{\varrho\sigma\sigma'}$	x, y, z
$B_{1/2g}$	1	1	-1	-1			
$B_{1/2u}$	1	-1	-1	1			

TABLE A-II

Point Groups C_{2h} and C_{3h}

C_{2h}	E	C_2	σ_h	i	R	iR	Raman activity			Dipole moment
A_g	1	1	1	1	1	1	$\alpha_{xx}, \alpha_{yy}, \alpha_{zz}, \alpha_{xy} \pm \alpha_{yx}$	$\alpha_0^0, \alpha_0^1, \alpha_2^2, \alpha_0^2, \alpha_{-2}^2$		
A_u	1	1	-1	-1	1	-1			$\beta_{zzz}, \beta_{yyz}, \beta_{zxx}, \beta_{xyz}$	z
B_g	1	-1	-1	1	1	1	$\alpha_{zy} \pm \alpha_{yz}, \alpha_{xz} \pm \alpha_{zx}$	$\alpha_1^1, \alpha_{-1}^1, \alpha_1^2, \alpha_{-1}^2$		
B_u	1	-1	1	-1	1	-1			$\beta_{xxx}, \beta_{yyy}, \beta_{xxy}, \beta_{zzx}, \beta_{xyy}, \beta_{yzz}$	x, y
$E_{1/2g}$	2	0	0	2	-2	-2				
$E_{1/2u}$	2	0	0	-2	-2	2				

C_{3h}	E	$2C_3$	σ_h	$2S_3$	R	$2C_3^2$	$2S_3^5$	Raman activity			Dipole moment
A'	1	1	1	1	1	1	1	$\alpha_{xx} + \alpha_{yy}, \alpha_{zz}, \alpha_{xy} - \alpha_{yx}$	$\alpha_0^0, \alpha_0^1, \alpha_0^2$	$\beta_{xxx} - 3\beta_{xyy}, \beta_{yyy} - 3\beta_{xxy}$	
A''	1	1	-1	-1	1	1	-1			$\beta_{yyz} + \beta_{zxx}, \beta_{zzz}$	z
E'	2	-1	2	-1	2	-1	-1	$(\alpha_{xx} - \alpha_{yy}, \alpha_{xy} + \alpha_{yx})$	$\alpha_2^2, \alpha_{-2}^2$	$(\beta_{xxx} + \beta_{xyy}, \beta_{yyy} + \beta_{xxy})$,	(x, y)
E''	2	-1	-2	1	2	-1	1	$(\alpha_{zy} \pm \alpha_{yz}, \alpha_{zx} \pm \alpha_{xz})$	$\alpha_1^2, \alpha_{-1}^2, \alpha_1^1, \alpha_{-1}^1$	β_{zzx}, β_{yzz}	
$E_{1/2}$	2	1	0	$\sqrt{3}$	-2	-1	$-\sqrt{3}$				
$E_{3/2}$	2	-2	0	0	-2	2	0				
$E_{5/2}$	2	1	0	$-\sqrt{3}$	-2	-1	$\sqrt{3}$				

TABLE A-III

Point Groups D_2, C_{2v} and D_{2h} ($D_{2h} = D_2 \times C_i$)

D_2 / C_{2v}	E / E	$C_2(z)$ / $C_2(z)$	$C_2(y)$ / $\sigma_v(xz)$	$C_2(x)$ / $\sigma_v(yz)$	R / R	Raman activity			Dipole moment
A A_1	1	1	1	1	1	$\alpha_{xx}, \alpha_{yy}, \alpha_{zz}$	$\alpha_0^0, \alpha_0^2, \alpha_2^2 + \alpha_{-2}^2$ (B_1 for C_{2v})	β_{xyz} (A_2 for C_{2v})	
B_1 A_2	1	1	−1	−1	1	$\alpha_{yx} \pm \alpha_{xy}$	α_0^1	$\beta_{zzz}, \beta_{yyz}, \beta_{zxx}$ (A_1 for C_{2v})	z (A_1 for C_{2v})
B_2 B_1	1	−1	1	−1	1	$\alpha_{zx} \pm \alpha_{xz}$	$\alpha_1^1 + \alpha_{-1}^1, \alpha_1^2 - \alpha_{-1}^2,$ $\alpha_2^2 - \alpha_{-2}^2$	$\beta_{yyy}, \beta_{xxy}, \beta_{yzz}$ (B_2 for C_{2v})	y (B_2 for C_{2v})
B_3 B_2	1	−1	−1	1	1	$\alpha_{yz} \pm \alpha_{zy}$	$\alpha_1^1 - \alpha_{-1}^1, \alpha_1^2 + \alpha_{-1}^2$	$\beta_{xxx}, \beta_{zzx}, \beta_{xyy}$ (B_1 for C_{2v})	x (B_1 for C_{2v})
$E_{1/2}$ $E_{1/2}$	2	0	0	0	−2				

TABLE A-IV

Point Groups D_3, C_{3v} and D_{3d} ($D_{3d} = D_3 \times C_i$)

D_3 / C_{3v}	E / E	$2C_3$ / $2C_3$	$3C_2$ / $3\sigma_v$	R / R	$2C_3^2$ / $2C_3^2$	Raman activity			Dipole moment
A_1	1	1	1	1	1	$\alpha_{xx} + \alpha_{yy}, \alpha_{zz}$	α_0^0, α_0^2	$\beta_{xxx} - 3\beta_{xyy}$	
A_2	1	1	−1	1	1	$\alpha_{xy} - \alpha_{yx}$	α_0^1	$\beta_{yyy} - 3\beta_{xxy}, \beta_{yyz} + \beta_{zxx}$ (A_1 for C_{3v}), β_{zzz} (A_1 for C_{3v})	z (A_1 for C_{3v})
E	2	−1	0	2	−1	$(\alpha_{xx} - \alpha_{yy}, \alpha_{xy} + \alpha_{yx}),$ $(\alpha_{yz} \pm \alpha_{zy}, \alpha_{zx} \pm \alpha_{xz})$	$\alpha_1^1, \alpha_{-1}^1, \alpha_2^2, \alpha_1^2, \alpha_{-1}^2, \alpha_{-2}^2$	$(\beta_{xxx} + \beta_{xyy}, \beta_{yyy} + \beta_{xxy}),$ $(\beta_{yyz} - \beta_{zxx}, \beta_{xyz}), (\beta_{zzx}, \beta_{yzz})$	(x, y)
$E_{1/2}$	2	1	0	−2	−1				
$E_{3/2}$	2	−2	0	−2	2				

TABLE A-V

Point groups D_4, C_{4v}, D_{2d} and $D_{4h} (D_{4h} = D_4 \times C_i)$

D_4 C_{4v} D_{2d}	E E E	$2C_4(z)$ $2C_4(z)$ $2S_4(z)$	$C_4^2 \equiv C''_2$ $C_4^2 \equiv C''_2$ $S_4^2 \equiv C''_2$	$2C_2$ $2\sigma_v$ $2C_2$	$2C'_2$ $2\sigma_d$ $2\sigma_d$	R R R	$2C_4^3$ $2C_4^3$ $2S_4^3$	Raman activity			Dipole moment
A_1	1	1	1	1	1	1	1	$\alpha_{xx} + \alpha_{yy}, \alpha_{zz}$	α_0^0, α_0^2		
A_2	1	1	1	-1	-1	1	1	$\alpha_{xy} - \alpha_{yx}$	α_0^1	$\beta_{yyz} + \beta_{zxx}$ (A_1 for C_{4v}, B_2 for D_{2d}), β_{zzz} (A_1 for C_{4v}, B_2 for D_{2d})	z (A_1 for C_{4v}, B_2 for D_{2d})
B_1	1	-1	1	1	-1	1	-1	$\alpha_{xx} - \alpha_{yy}$	$\alpha_2^2 + \alpha_{-2}^2$	β_{xyz} (B_2 for C_{4v}, A_1 for D_{2d})	
B_2	1	-1	1	-1	1	1	-1	$\alpha_{xy} + \alpha_{yx}$	$\alpha_2^2 - \alpha_{-2}^2$	$\beta_{yyz} - \beta_{zxx}$ (B_1 for C_{4v}, A_2 for D_{2d})	
E	2	0	-2	0	0	2	0	$(\alpha_{yz} \pm \alpha_{zy}, \alpha_{zx} \pm \alpha_{xz})$	$\alpha_1^1, \alpha_{-1}^1, \alpha_1^2, \alpha_{-1}^2$	$(\beta_{xxx}, \beta_{yyy}), (\beta_{xyy}, \beta_{xxy}), (\beta_{zzx}, \beta_{yzz})$	(x, y)
$E_{1/2}$	2	$+\sqrt{2}$	0	0	0	-2	$-\sqrt{2}$				
$E_{3/2}$	2	$-\sqrt{2}$	0	0	0	-2	$+\sqrt{2}$				

TABLE A-VI

Point groups D_5, C_{5v} and D_{5d} $(D_{5d} = D_5 \times C_i)$

D_5 C_{5v}	E E	$2C_5$ $2C_5$	$2C_5^2$ $2C_5^2$	$5C_2$ $5\sigma_v$	R R	$2C_5^4$ $2C_5^4$	$2C_5^3$ $2C_5^3$	Raman activity			Dipole moment
A_1	1	1	1	1	1	1	1	$\alpha_{xx} + \alpha_{yy}, \alpha_{zz}$	α_0^0, α_0^2		
A_2	1	1	1	-1	1	1	1	$\alpha_{xy} - \alpha_{yx}$	α_0^1	$\beta_{yyz} + \beta_{zxx}$ (A_1 for C_{5v}), β_{zzz} (A_1 for C_{5v})	z (A_1 for C_{5v})
E_1	2	$\frac{-1+\sqrt{5}}{2}$	$\frac{-1-\sqrt{5}}{2}$	0	2	$\frac{-1+\sqrt{5}}{2}$	$\frac{-1-\sqrt{5}}{2}$	$(\alpha_{yz} \pm \alpha_{zy}, \alpha_{zx} \pm \alpha_{xz})$	$\alpha_1^1, \alpha_{-1}^1, \alpha_1^2, \alpha_{-1}^2$	$(\beta_{xxx} + \beta_{xyy}, \beta_{yyy} + \beta_{xxy}), (\beta_{zzx}, \beta_{yzz})$	
E_2	2	$\frac{-1-\sqrt{5}}{2}$	$\frac{-1+\sqrt{5}}{2}$	0	2	$\frac{-1-\sqrt{5}}{2}$	$\frac{-1+\sqrt{5}}{2}$	$(\alpha_{xx} - \alpha_{yy}, \alpha_{xy} + \alpha_{yx})$	$\alpha_2^2, \alpha_{-2}^2$	$(\beta_{xxx} - 3\beta_{xyy}, \beta_{yyy} - 3\beta_{xxy}), (\beta_{yyz} - \beta_{zxx}, \beta_{xyz})$	(x, y)
$E_{1/2}$	2	$\frac{1+\sqrt{5}}{2}$	$\frac{-1+\sqrt{5}}{2}$	0	-2	$\frac{-1-\sqrt{5}}{2}$	$\frac{1-\sqrt{5}}{2}$				
$E_{3/2}$	2	$\frac{1-\sqrt{5}}{2}$	$\frac{-1-\sqrt{5}}{2}$	0	-2	$\frac{-1+\sqrt{5}}{2}$	$\frac{1+\sqrt{5}}{2}$				
$E_{5/2}$	2	-2	2	0	-2	$+2$	-2				

TABLE A-VII

Point groups D_{3h}, D_{6i} C_{6v} and D_{6h} ($D_{6h} = D_6 \times C_i$)

D_6 / C_{6v} / D_{3h}	E / E / E	$2C_6(z)$ / $2C_6(z)$ / $2S_3(z)$	$2C_6^2 \equiv 2C_3$ / $2C_6^2 \equiv 2C_3$ / $2C_3(z)$	$C_6^3 \equiv C'_2$ / $C_6^3 \equiv C''_2$ / σ_h	$3C'_2$ / $3\sigma_v$ / $3C_2$	$3C''_2$ / $3\sigma_d$ / $3\sigma_v$	R / R / R	$2C_6^5$ / $2C_6^5$ / $2S_3^5$	$2C_6^4 \equiv 2C_3^2$ / $2C_3^2$ / $2C_3^2$	Raman activity			Dipole moment
A'_1 A_1	1	1	1	1	1	1	1	1	1	$\alpha_{xx}+\alpha_{yy}$, α_{zz}	α_0^0, α_0^2	$\beta_{xxx}-3\beta_{xyy}$ (A_2 for D_6, B_1 for C_{6v})	
A'_2 A_2	1	1	1	1	−1	−1	1	1	1	$\alpha_{xy}-\alpha_{yx}$	α_0^1	$\beta_{yyy}-3\beta_{xxy}$ (B_2 for D_6, B_2 for C_{6v})	
A''_1 B_1	1	−1	1	−1	1	−1	1	−1	1				z (A_2 for D_6 A_1 for C_{6v})
A''_2 B_2	1	−1	1	−1	−1	1	1	−1	1			$\beta_{yyz}+\beta_{zxx}$, β_{zzz} (A_2 for D_6, A_1 for C_{6v})	
E'' E_1	2	1	−1	−2	0	0	2	1	−1	($\alpha_{yz}\pm\alpha_{zy}$, $\alpha_{zx}\pm\alpha_{xz}$)	α_1^1, α_{-1}^1, α_1^2, α_{-1}^2	($\beta_{yyz}-\beta_{zxx}$, β_{xyz}) E_2 for D_6 and C_{6v}	
E' E_2	2	−1	−1	2	0	0	2	−1	−1	($\alpha_{xx}-\alpha_{yy}$, $\alpha_{xy}+\alpha_{yx}$)	α_2^2, α_{-2}^2	($\beta_{xxx}+\beta_{xyy}$, $\beta_{yyy}+\beta_{xxy}$), (β_{zzx}, β_{yzz}) E_1 for D_6 and C_{6v}	(x, y) (E_1 for D_6 and C_{6v})
$E_{1/2}E_{1/2}$	2	$\sqrt{3}$	1	0	0	0	−2	$-\sqrt{3}$	−1				
$E_{3/2}E_{3/2}$	2	0	−2	0	0	0	−2	0	2				
$E_{5/2}E_{5/2}$	2	$-\sqrt{3}$	1	0	0	0	−2	$\sqrt{3}$	−1				

TABLE A-VIII
Point group D_{4d}

D_{4d}	E	$2S_8(z)$	$2S_8^2 = 2C_4$	$2S_8^3$	$S_8^4 = C_2$	$4C_2$	$4\sigma_d$	R	$2S_8R$	$2C_4R$	$2S_8^3R$	Raman activity			Dipole moment
1	1	1	1	1	1	+1	[illegible]	1	1	1	1	$\alpha_{xx}+\alpha_{yy}$, α_{zz}	α_0^0, α_0^2		
2	1	1	1	1	1	−1	−1	1	1	1	1	$\alpha_{xy}-\alpha_{yx}$	α_0^1		
1	1	−1	1	−1	1	+1	−1	1	−1	1	−1				
2	1	−1	1	−1	1	−1	+1	1	−1	1	−1			$\beta_{yyz}+\beta_{zxx}$, β_{zzz}	z
1	2	$\sqrt{2}$	0	$-\sqrt{2}$	−2	0	0	2	$\sqrt{2}$	0	$-\sqrt{2}$			$(\beta_{xxx}+\beta_{xyy},$ $\beta_{yyy}+\beta_{xxy})$ $(\beta_{zzx}, \beta_{yzz})$	(x, y)
2	2	0	−2	0	+2	0	0	2	0	−2	0	$(\alpha_{xx}-\alpha_{yy},$ $\alpha_{xy}+\alpha_{yx})$	α_2^2, α_{-2}^2	$(\beta_{yyz}-\beta_{zxx},$ $\beta_{xyz})$	
3	2	$-\sqrt{2}$	0	$\sqrt{2}$	−2	0	0	2	$-\sqrt{2}$	0	$\sqrt{2}$	$(\alpha_{yz}\pm\alpha_{zy},$ $\alpha_{zx}\pm\alpha_{xz})$	α_1^1, α_{-1}^1, α_1^2, α_{-1}^2	$(\beta_{xxx}-3\beta_{xyy},$ $\beta_{yyy}-3\beta_{xxy})$	
1/2	2	$\sqrt{2+\sqrt{2}}$	$\sqrt{2}$	$\sqrt{2-\sqrt{2}}$	0	0	0	−2	$-\sqrt{2+\sqrt{2}}$	$-\sqrt{2}$	$-\sqrt{2-\sqrt{2}}$				
3/2	2	$\sqrt{2-\sqrt{2}}$	$-\sqrt{2}$	$-\sqrt{2+\sqrt{2}}$	0	0	0	−2	$-\sqrt{2-\sqrt{2}}$	$+\sqrt{2}$	$\sqrt{2+\sqrt{2}}$				
5/2	2	$-\sqrt{2-\sqrt{2}}$	$-\sqrt{2}$	$\sqrt{2+\sqrt{2}}$	0	0	0	−2	$\sqrt{2-\sqrt{2}}$	$+\sqrt{2}$	$-\sqrt{2+\sqrt{2}}$				
7/2	2	$-\sqrt{2+\sqrt{2}}$	$\sqrt{2}$	$-\sqrt{2-\sqrt{2}}$	0	0	0	−2	$\sqrt{2+\sqrt{2}}$	$-\sqrt{2}$	$\sqrt{2-\sqrt{2}}$				

TABLE A-IX

Point groups O, T_d and O_h ($O_h = O \times C_i$).

O[a]		E	$8C_3$	$6C_2$	$6C_4$	$3C_4^2 \equiv 3C''_2$	R	$8C_3^2$	$6C_4^3$	Raman activity			
T_d		E	$8C_3$	$6\sigma_d$	$6S_4$	$3S_4^2 \equiv 3C_2$	R	$8C_3^2$	$6S_4^3$				
A_1	Γ_1	1	1	1	1	1	1	1	1	$\alpha_{xx}+\alpha_{yy}+\alpha_{zz}$	α_0^0		
A_2	Γ_2	1	1	−1	−1	1	1	1	−1			β_{xyz} (A_1 for T_d)	
E	Γ_3	2	−1	0	0	2	2	−1	0	$(\sqrt{3}\alpha_{xx}-\sqrt{3}\alpha_{yy}$, $\alpha_{xx}+\alpha_{yy}-2\alpha_{zz})$	α_0^2, $\alpha_2^2+\alpha_2^2$		
$F_1(T_1)$	Γ_4	3	0	−1	1	−1	3	0	1		α_1^1, α_0^1, α_{-1}^1	$(\beta_{xxx}, \beta_{yyy}, \beta_{zzz})$, $(\beta_{xyy}+\beta_{zzx}, \beta_{yzz}+\beta_{xxy})$, $(\beta_{zxx}+\beta_{yyz})F_2$ for T_d	(x, y, z) F_2 for T_d
$F_2(T_2)$	Γ_5	3	0	1	−1	−1	3	0	−1	$(\alpha_{xy}\pm\alpha_{yx}$, $\alpha_{xz}\pm\alpha_{zx}$, $\alpha_{yz}\pm\alpha_{zy})$	$\alpha_1^2, \alpha_{-1}^2, \alpha_2^2-\alpha_{-2}^2$	$(\beta_{xyy}-\beta_{zzx}, \beta_{yzz}-\beta_{xxy}$, $\beta_{zxx}-\beta_{yyz})$ F_1 for T_d	
$E_{1/2}$	Γ_6	2	1	0	$\sqrt{2}$	0	−2	−1	$-\sqrt{2}$				
$E_{5/2}$	Γ_7	2	1	0	$-\sqrt{2}$	0	−2	−1	$\sqrt{2}$				
$G_{3/2}$	Γ_8	4	−1	0	0	0	−4	1	0				

[a] x-axis along C_2.

TABLE A-X

Point groups D_∞, $C_{\infty v}$ and $D_{\infty h}$ ($D_{\infty h} = C_{\infty v} \times C_i$)

D_∞	E	$2C_\infty^{\varphi}$	$2C_\infty^{2\varphi}$	$2C_\infty^{3\varphi}$	...	∞C_2	R	$2C_\infty^{\varphi}R$	...	Raman activity			Dipole moment
$C_{\infty v}$	E	$2C_\infty^{\varphi}$	$2C_\infty^{2\varphi}$	$2C_\infty^{3\varphi}$	...	$\infty\sigma_v$	R	$2C_\infty^{\varphi}R$	...				
Σ^+	1	1	1	1		1	1	1		$\alpha_{xx}+\alpha_{yy}, \alpha_{zz}$	α_0^0, α_0^2	$\beta_{yyz}+\beta_{zxx}$, β_{zzz}	z
Σ^-	1	1	1	1		-1	1	1		$\alpha_{xy}-\alpha_{yx}$	α_0^1		
Π	$+2$	$2\cos\varphi$	$2\cos 2\varphi$	$2\cos 3\varphi$	...	0	$+2$	$2\cos\varphi$	...	$(\alpha_{yz}, \pm\alpha_{zy},$ $\alpha_{zx}\pm\alpha_{xz})$	$\alpha_1^1, \alpha_{-1}^1, \alpha_1^2,$ α_{-1}^2	$(\beta_{xxx}+\beta_{xyy}, \beta_{yyy}+\beta_{xxy})$, $(\beta_{zzx}, \beta_{yzz})$	(x, y)
Δ	$+2$	$2\cos 2\varphi$	$2\cos 2\cdot 2\varphi$	$2\cos 3\cdot 2\varphi$	...	0	$+2$	$2\cos 2\varphi$	...	$(\alpha_{xx}-\alpha_{yy},$ $\alpha_{xy}+\alpha_{yx})$	α_2^2, α_{-2}^2	$(\beta_{yyz}-\beta_{zxx}, \beta_{xyz})$	
Φ	$+2$	$2\cos 3\varphi$	$2\cos 2\cdot 3\varphi$	$2\cos 3\cdot 3\varphi$	...	0	$+2$	$2\cos 3\varphi$	...			$(\beta_{xxx}-3\beta_{xyz}, \beta_{yyy}-3\beta_{xxy})$	
...	...	...	...	...	...	...	...	...	...				
$E_{1/2}$	$+2$	$2\cos\frac{1}{2}\varphi$	$2\cos\varphi$	$2\cos 3\times\frac{1}{2}\varphi$	...	0	-2	$-2\cos\frac{1}{2}\varphi$	...				
$E_{3/2}$	$+2$	$2\cos\frac{3}{2}\varphi$	$2\cos 3\varphi$	$2\cos 3\times\frac{3}{2}\varphi$	...	0	-2	$-2\cos\frac{3}{2}\varphi$	...				
$E_{5/2}$	$+2$	$2\cos\frac{5}{2}\varphi$	$2\cos 5\varphi$	$2\cos 3\times\frac{5}{2}\varphi$	...	0	-2	$-2\cos\frac{5}{2}\varphi$	...				
...	...	...	...	...	...	...	...	...	...				

For scattering processes which follow Placzek's polarizability theory the antisymmetric tensors should be omitted. In the third column are the elements of the hyper-polarizability tensor. The tensor operator $\hat{\alpha}$ is even and $\hat{\beta}$ is odd. Thus for groups with a center of inversion α belongs always to the gerade species and β to the ungerade species. The latter condition pertains also to the electric dipole operator. The elements of α are under certain conditions symmetric in the indices ϱ and σ, for the elements of the hyper-polarizability tensor $\beta_{\varrho\sigma\sigma'}$, the indices ϱ and σ' may be interchanged. In order to keep the number of elements to a reasonable size, this particular aspect is only shown for α.

In order to establish the selection rules for the Raman effect it is necessary to have at hand tables in which the direct and also the antisymmetric product of representations are given. The information is contained in Table A-XI given below. Only for the electronic Raman effect and for the other scattering processes which involve degenerate electronic states should the antisymmetric product be considered. Also the species of the extended groups should be omitted for the normal and stimulated and hyper vibrational and rotational Raman effects.

TABLE A-XI

Direct products of representations (species) for some important point groups
Species to be omitted in the symmetrical product of a degenerate species with itself are put in square brackets. They represent the antisymmetric product.

C_s

	A'	A''	$E_{\frac{1}{2}}$
A'	A'	A''	$E_{\frac{1}{2}}$
A''		A'	$E_{\frac{1}{2}}$
$E_{\frac{1}{2}}$			$[A']$, A', A'', A''

C_i, (C_1)[a]

	A_g	A_u	$B_{\frac{1}{2}g}$	$B_{\frac{1}{2}u}$
A_g	A_g	A_u	$B_{\frac{1}{2}g}$	$B_{\frac{1}{2}u}$
A_u		A_g	$B_{\frac{1}{2}u}$	$B_{\frac{1}{2}g}$
$B_{\frac{1}{2}g}$			A_g	A_u
$B_{\frac{1}{2}u}$				A_g

C_{2v}, (C_2)[b], (C_{2h})[b, c]

	A_1	A_2	B_1	B_2	$E_{\frac{1}{2}}$
A_1	A_1	A_2	B_1	B_2	$E_{\frac{1}{2}}$
A_2		A_1	B_2	B_1	$E_{\frac{1}{2}}$
B_1			A_1	A_2	$E_{\frac{1}{2}}$
B_2				A_1	$E_{\frac{1}{2}}$
$E_{\frac{1}{2}}$					$[A_1]$, A_2, B_1, B_2

C_3

	A	E	$E_{\frac{1}{2}}$	$B_{\frac{3}{2}}$
A	A	E	$E_{\frac{1}{2}}$	$B_{\frac{3}{2}}$
E		$[A]$, A, E	$E_{\frac{1}{2}}$, $2B_{\frac{3}{2}}$	$E_{\frac{1}{2}}$
$E_{\frac{1}{2}}$			$[A]$, A, E	E
$B_{\frac{3}{2}}$				A

D_2, (D_{2h})[c]

	A	B_1	B_2	B_3	$E_{\frac{1}{2}}$
A	A	B_1	B_2	B_3	$E_{\frac{1}{2}}$
B_1		A	B_3	B_2	$E_{\frac{1}{2}}$
B_2			A	B_1	$E_{\frac{1}{2}}$
B_3				A	$E_{\frac{1}{2}}$
$E_{\frac{1}{2}}$					$[A]$, B_1, B_2, B_3

D_3, C_{3v}, (D_{3d})[c]

	A_1	A_2	E	$E_{\frac{1}{2}}$	$E_{\frac{3}{2}}$
A_1	A_1	A_2	E	$E_{\frac{1}{2}}$	$E_{\frac{3}{2}}$
A_2		A_1	E	$E_{\frac{1}{2}}$	$E_{\frac{3}{2}}$
E			A_1, $[A_2]$, E	$E_{\frac{1}{2}}$, $E_{\frac{3}{2}}$	$E_{\frac{1}{2}}$, $E_{\frac{1}{2}}$
$E_{\frac{1}{2}}$				$[A_1]$, A_2, E	E, E
$E_{\frac{3}{2}}$					$[A_1]$, A_2, A_1, A_2

Table A-XI (Continued)

D_{3h}, (C_{3h})[b]

	A'_1	A'_2	A''_1	A''_2	E'	E''	$E_{1/2}$	$E_{3/2}$	$E_{5/2}$
A'_1	A'_1	A'_2	A''_1	A''_2	E'	E''	$E_{1/2}$	$E_{3/2}$	$E_{5/2}$
A'_2		A'_1	A''_2	A''_1	E'	E''	$E_{1/2}$	$E_{3/2}$	$E_{5/2}$
A''_1			A'_1	A'_2	E''	E'	$E_{5/2}$	$E_{3/2}$	$E_{1/2}$
A''_2				A'_1	E''	E'	$E_{5/2}$	$E_{3/2}$	$E_{1/2}$
E'					$A'_1,[A'_2],E'$	A''_1,A''_2,E''	$E_{3/2},E_{5/2}$	$E_{1/2},E_{5/2}$	$E_{1/2},E_{3/2}$
E''						$A'_1,[A'_2],E'$	$E_{3/2},E_{1/2}$	$E_{1/2},E_{5/2}$	$E_{3/2},E_{5/2}$
$E_{1/2}$							$[A'_1],A'_2,E''$	E',E''	A_1'',A_2'',E'
$E_{3/2}$								$[A'_1],A'_2,A''_1,A''_2$	E',E''
$E_{5/2}$									$[A'_1],A'_2,E''$

D_4, C_{4v}, D_{2d}, (D_{4h})[c]

	A_1	A_2	B_1	B_2	E	$E_{1/2}$	$E_{3/2}$
A_1	A_1	A_2	B_1	B_2	E	$E_{1/2}$	$E_{3/2}$
A_2		A_1	B_2	B_1	E	$E_{1/2}$	$E_{3/2}$
B_1			A_1	A_2	E	$E_{3/2}$	$E_{1/2}$
B_2				A_1	E	$E_{3/2}$	$E_{1/2}$
E					$A_1, [A_2], B_1, B_2$	$E_{1/2}, E_{3/2}$	$E_{1/2}, E_{3/2}$
$E_{1/2}$						$[A_1], A_2, E$	B_1, B_2, E
$E_{3/2}$							$[A_1], A_2, E$

D_5, C_{5v}, (D_{5d})[c]

	A_1	A_2	E_1	E_2	$E_{1/2}$	$E_{3/2}$	$E_{5/2}$
A_1	A_1	A_2	E_1	E_2	$E_{1/2}$	$E_{3/2}$	$E_{5/2}$
A_2		A_1	E_1	E_2	$E_{1/2}$	$E_{3/2}$	$E_{5/2}$
E_1			$A_1, [A_2], E_2$	E_1, E_2	$E_{1/2}, E_{3/2}$	$E_{1/2}, E_{5/2}$	$E_{3/2}, E_{3/2}$
E_2				$A_1, [A_2], E_1$	$E_{3/2}, E_{5/2}$	$E_{1/2}, E_{5/2}$	$E_{3/2}, E_{3/2}$
$E_{1/2}$					$[A_1], A_2, E_1$	E_1, E_2	E_2, E_2
$E_{3/2}$						$[A_1], A_2, E_2$	E_1, E_1
$E_{5/2}$							$[A_1], A_2, A_1, A_2$

D_6, C_{6v}, (D_{6h})[c]

	A_1	A_2	B_1	B_2	E_1	E_2	$E_{1/2}$	$E_{3/2}$	$E_{5/2}$
A_1	A_1	A_2	B_1	B_2	E_1	E_2	$E_{1/2}$	$E_{3/2}$	$E_{5/2}$
A_2		A_1	B_2	B_1	E_1	E_2	$E_{1/2}$	$E_{3/2}$	$E_{5/2}$
B_1			A_1	A_2	E_2	E_1	$E_{5/2}$	$E_{3/2}$	$E_{1/2}$
B_2				A_1	E_2	E_1	$E_{5/2}$	$E_{3/2}$	$E_{1/2}$
E_1					$A_1, [A_2], E_2$	B_1, B_2, E_1	$E_{1/2}, E_{3/2}$	$E_{1/2}, E_{5/2}$	$E_{3/2}, E_{5/2}$
E_2						$A_1, [A_2], E_2$	$E_{3/2}, E_{5/2}$	$E_{1/2}, E_{5/2}$	$E_{1/2}, E_{3/2}$
$E_{1/2}$							$[A_1], A_2, E_1$	E_1, E_2	B_1, B_2, E_2
$E_{3/2}$								$[A_1], A_2, B_1, B_2$	E_1, E_2
$E_{5/2}$									$[A_1]A_2, E_1$

Table A-XI (Continued)

D_{4d}

	A_1	A_2	B_1	B_2	E_1	E_2	E_3	$E_{\frac{1}{2}}$	$E_{\frac{3}{2}}$	$E_{\frac{5}{2}}$	$E_{\frac{7}{2}}$
A_1	A_1	A_2	B_1	B_2	E_1	E_2	E_3	$E_{\frac{1}{2}}$	$E_{\frac{3}{2}}$	$E_{\frac{5}{2}}$	$E_{\frac{7}{2}}$
A_2		A_1	B_2	B_1	E_1	E_2	E_3	$E_{\frac{1}{2}}$	$E_{\frac{3}{2}}$	$E_{\frac{5}{2}}$	$E_{\frac{7}{2}}$
B_1			A_1	A_2	E_3	E_2	E_1	$E_{\frac{7}{2}}$	$E_{\frac{5}{2}}$	$E_{\frac{3}{2}}$	$E_{\frac{1}{2}}$
B_2				A_1	E_3	E_2	E_1	$E_{\frac{7}{2}}$	$E_{\frac{5}{2}}$	$E_{\frac{3}{2}}$	$E_{\frac{1}{2}}$
E_1					A_1, [A_2], E_2	E_1, E_3	B_1, B_2, E_2	$E_{\frac{1}{2}}$, $E_{\frac{3}{2}}$	$E_{\frac{1}{2}}$, $E_{\frac{5}{2}}$	$E_{\frac{3}{2}}$, $E_{\frac{7}{2}}$	$E_{\frac{5}{2}}$, $E_{\frac{7}{2}}$
E_2						A_1, [A_2], B_1, B_2	E_1, E_3	$E_{\frac{3}{2}}$, $E_{\frac{5}{2}}$	$E_{\frac{1}{2}}$, $E_{\frac{7}{2}}$	$E_{\frac{1}{2}}$, $E_{\frac{7}{2}}$	$E_{\frac{3}{2}}$, $E_{\frac{5}{2}}$
E_3							A_1, [A_2], E_2	$E_{\frac{5}{2}}$, $E_{\frac{7}{2}}$	$E_{\frac{3}{2}}$, $E_{\frac{7}{2}}$	$E_{\frac{1}{2}}$, $E_{\frac{5}{2}}$	$E_{\frac{1}{2}}$, $E_{\frac{3}{2}}$
$E_{\frac{1}{2}}$								[A_1], A_2, E_1	E_1, E_2	E_2, E_3	B_1, B_2, E_3
$E_{\frac{3}{2}}$									[A_1], A_2, E_3	B_1, B_2, E_1	E_2, E_3
$E_{\frac{5}{2}}$										[A_1], A_2, E_3	E_1, E_2
$E_{\frac{7}{2}}$											[A_1], A_2, E_1

D_∞, $C_{\infty v}$, $(D_{\infty h})$ [b]

	Σ^+	Σ^-	Π	Δ	Φ	…	$E_{\frac{1}{2}}$	$E_{\frac{3}{2}}$	$E_{\frac{5}{2}}$	…
Σ^+	Σ^+	Σ^-	Π	Δ	Φ	…	$E_{\frac{1}{2}}$	$E_{\frac{3}{2}}$	$E_{\frac{5}{2}}$	…
Σ^-		Σ^+	Π	Δ	Φ	…	$E_{\frac{1}{2}}$	$E_{\frac{3}{2}}$	$E_{\frac{5}{2}}$	…
Π			Σ^+, [Σ^-], Δ	Π, Φ	Δ, Γ	…	$E_{\frac{1}{2}}$, $E_{\frac{3}{2}}$	$E_{\frac{1}{2}}$, $E_{\frac{5}{2}}$	$E_{\frac{3}{2}}$, $E_{\frac{7}{2}}$	…
Δ				Σ^+, [Σ^-], Γ	Π, H	…	$E_{\frac{3}{2}}$, $E_{\frac{5}{2}}$	$E_{\frac{1}{2}}$, $E_{\frac{7}{2}}$	$E_{\frac{1}{2}}$, $E_{\frac{9}{2}}$	…
Φ					Σ^+, [Σ^-], I	…	$E_{\frac{5}{2}}$, $E_{\frac{7}{2}}$	$E_{\frac{3}{2}}$, $E_{\frac{9}{2}}$	$E_{\frac{1}{2}}$, $E_{\frac{11}{2}}$	…
…	…	…	…	…	…	…	…	…	…	…
$E_{\frac{1}{2}}$							[Σ^+], Σ^-, Π	Π, Δ	Δ, Φ	…
$E_{\frac{3}{2}}$								[Σ^+], Σ^-, Φ	Π, Γ	…
$E_{\frac{5}{2}}$									[Σ^+], Σ^-, H	…
…	…	…	…	…	…	…	…	…	…	…

O, T_d, (O_h) [b]

	A_1	A_2	E	F_1	F_2	$E_{\frac{1}{2}}$	$E_{\frac{5}{2}}$	$G_{\frac{3}{2}}$
A_1	A_1	A_2	E	F_1	F_2	$E_{\frac{1}{2}}$	$E_{\frac{5}{2}}$	$G_{\frac{3}{2}}$
A_2		A_1	E	F_2	F_1	$E_{\frac{5}{2}}$	$E_{\frac{1}{2}}$	$G_{\frac{3}{2}}$
E			A_1, [A_2], E	F_1, F_2	F_1, F_2	$G_{\frac{3}{2}}$	$G_{\frac{3}{2}}$	$E_{\frac{1}{2}}$, $E_{\frac{5}{2}}$, $G_{\frac{3}{2}}$
F_1				A_1, E, [F_1,] F_2	A_2, E, F_1, F_2	$E_{\frac{1}{2}}$, $G_{\frac{3}{2}}$	$E_{\frac{5}{2}}$, $G_{\frac{3}{2}}$	$E_{\frac{1}{2}}$, $E_{\frac{5}{2}}$, $2G_{\frac{3}{2}}$
F_2					A_1, E, [F_1], F_2	$E_{\frac{5}{2}}$, $G_{\frac{3}{2}}$	$E_{\frac{1}{2}}$, $G_{\frac{3}{2}}$	$E_{\frac{1}{2}}$, $E_{\frac{5}{2}}$, $2G_{\frac{3}{2}}$
$E_{\frac{1}{2}}$						[A_1], F_1	A_2, F_2	E, F_1, F_2
$E_{\frac{5}{2}}$							[A_1], F_1	E, F_1, F_2
$G_{\frac{3}{2}}$								[A_1], A_2, [E], $2F_1$, [F_2], F_2

[a] For this point group g and u should be omitted.
[b] For these point groups the subscripts 1 and 2 should be dropped.
[c] For these point groups the (g, u) rule must be added, that is $g \times g = g$, $g \times u = u$, $u \times u = g$.

TABLE A-XII

Species of spin functions for the more important point groups

For the groups in ( ) the subscripts *g* and *u* should be omitted. The number in front of the species indicates the number of times the species occurs. Note that the species $B_{1/2}$ and $B_{3/2}$ do not occur singly but in pairs corresponding to Kramers doublets.

S	$C_i(C_1)$	C_s	$C_{2h}(C_2)$	C_{2v}	$D_{2h}(D_2)$
0	A_g	A'	A_g	A_1	A_g
$\frac{1}{2}$	$2B_{\frac{1}{2}g}$	$E_{\frac{1}{2}}$	$E_{\frac{1}{2}g}$	$E_{\frac{1}{2}}$	$E_{\frac{1}{2}g}$
1	$3A_g$	$A'+2A''$	A_g+2B_g	$A_2+B_1+B_2$	$B_{1g}+B_{2g}+B_{3g}$
$\frac{3}{2}$	$2B_{\frac{1}{2}g}+2B_{\frac{1}{2}g}$	$2E_{\frac{1}{2}}$	$2E_{\frac{1}{2}g}$	$2E_{\frac{1}{2}}$	$2E_{\frac{1}{2}g}$
2	$A_g+2A_g+2A_g$	$3A'+2A''$	$3A_g+2B_g$	$2A_1+A_2+B_1+B_2$	$2A_g+B_{1g}+B_{2g}+B_{3g}$
$\frac{5}{2}$	$2B_{\frac{1}{2}g}+2B_{\frac{1}{2}g}+2B_{\frac{1}{2}g}$	$3E_{\frac{1}{2}}$	$3E_{\frac{1}{2}g}$	$3E_{\frac{1}{2}}$	$3E_{\frac{1}{2}g}$

S	C_3	C_{3h}	C_{3v}	$D_{3d}(D_3)$	D_{3h}
0	A	A'	A_1	A_{1g}	A'_1
$\frac{1}{2}$	$E_{\frac{1}{2}}$	$E_{\frac{1}{2}}$	$E_{\frac{1}{2}}$	$E_{\frac{1}{2}g}$	$E_{\frac{1}{2}}$
1	$A+E$	$A'+E''$	A_2+E	$A_{2g}+E_g$	A'_2+E''
$\frac{3}{2}$	$2B_{\frac{3}{2}}+E_{\frac{1}{2}}$	$E_{\frac{1}{2}}+E_{\frac{3}{2}}$	$E_{\frac{1}{2}}+E_{\frac{3}{2}}$	$E_{\frac{1}{2}g}+E_{\frac{3}{2}g}$	$E_{\frac{1}{2}}+E_{\frac{3}{2}}$
2	$A+2E$	$A'+E'+E''$	A_1+2E	$A_{1g}+2E_g$	$A'_1+E'+E''$
$\frac{5}{2}$	$2B_{\frac{3}{2}}+2E_{\frac{1}{2}}$	$E_{\frac{1}{2}}+E_{\frac{3}{2}}+E_{\frac{5}{2}}$	$2E_{\frac{1}{2}}+E_{\frac{3}{2}}$	$2E_{\frac{1}{2}g}+E_{\frac{3}{2}g}$	$E_{\frac{1}{2}}+E_{\frac{3}{2}}+E_{\frac{5}{2}}$

S	C_{4v}, D_{2d}	$D_{4h}(D_4)$	D_{4d}	$D_{5d}(D_5, C_{5v})$
0	A_1	A_{1g}	A_1	A_{1g}
$\frac{1}{2}$	$E_{\frac{1}{2}}$	$E_{\frac{1}{2}g}$	$E_{\frac{1}{2}}$	$E_{\frac{1}{2}g}$
1	A_2+E	$A_{2g}+E_g$	A_2+E_1	$A_{2g}+E_{1g}$
$\frac{3}{2}$	$E_{\frac{1}{2}}+E_{\frac{3}{2}}$	$E_{\frac{1}{2}g}+E_{\frac{3}{2}g}$	$E_{\frac{1}{2}}+E_{\frac{3}{2}}$	$E_{\frac{1}{2}}+E_{\frac{3}{2}g}$
2	$A_1+B_1+B_2+E$	$A_{1g}+B_{1g}+B_{2g}+E_g$	$A_1+E_1+E_2$	$A_{1g}+E_{1g}+E_{2g}$
$\frac{5}{2}$	$E_{\frac{1}{2}}+2E_{\frac{3}{2}}$	$E_{\frac{1}{2}g}+2E_{\frac{3}{2}g}$	$E_{\frac{1}{2}}+E_{\frac{3}{2}}+E_{\frac{5}{2}}$	$E_{\frac{1}{2}g}+E_{\frac{3}{2}g}+E_{\frac{5}{2}g}$

S	$D_{6h}(D_6, C_{6v})$	$D_{\infty h}(C_{\infty v})$	$O_h(T_d)$
0	A_{1g}	Σ_g^+	A_{1g}
$\frac{1}{2}$	$E_{\frac{1}{2}g}$	$E_{\frac{1}{2}g}$	$E_{\frac{1}{2}g}$
1	$A_{2g}+E_{1g}$	$\Sigma_g^-+\Pi_g$	F_{1g}
$\frac{3}{2}$	$E_{\frac{1}{2}g}+E_{\frac{3}{2}g}$	$E_{\frac{1}{2}g}+E_{\frac{3}{2}g}$	$G_{\frac{3}{2}g}$
2	$A_{1g}+E_{1g}+E_{2g}$	$\Sigma_g^++\Pi_g+\Delta_g$	E_g+F_{2g}
$\frac{5}{2}$	$E_{\frac{1}{2}}+E_{\frac{3}{2}g}+E_{\frac{5}{2}g}$	$E_{\frac{1}{2}g}+E_{\frac{3}{2}g}+E_{\frac{5}{2}g}$	$E_{\frac{5}{2}}+G_{\frac{3}{2}g}$

Suggestion for Further Reading

Herzberg, G., *Molecular Spectra and Molecular Structure, III. Electronic Spectra and Electronic Structure of Polyatomic Molecules*, Van Nostrand Reinhold Comp.

INDEX

GPSR Compliance

The European Union's (EU) General Product Safety Regulation (GPSR) is a set of rules that requires consumer products to be safe and our obligations to ensure this.

If you have any concerns about our products, you can contact us on ProductSafety@springernature.com

In case Publisher is established outside the EU, the EU authorized representative is:

Springer Nature Customer Service Center GmbH
Europaplatz 3
69115 Heidelberg, Germany

Batch number: 10614746

Printed by Printforce, the Netherlands